The Big Splash

The Big Splash

*A Scientific Discovery That Revolutionizes
the Way We View the Origin of Life, the Water We
Drink, the Death of the Dinosaurs, the Creation of
the Oceans, the Nature of the Cosmos, and the
Very Future of the Earth Itself*

by

Dr. Louis A. Frank with Patrick Huyghe

A Birch Lane Press Book
Published by Carol Publishing Group

A Birch Lane Press Book
Published by Carol Publishing Group

Editorial Offices   Sales & Distribution Offices
600 Madison Avenue   120 Enterprise Avenue
New York, NY 10022   Secaucus, NJ 07094

In Canada: Musson Book Company
A division of General Publishing Co. Limited
Don Mills, Ontario

Manufactured in the United States of America

10   9   8   7   6   5   4   3   2   1

Library of Congress Cataloging-in-Publication Data

Frank, Louis A.
    The big splash : a scientific discovery that revolutionizes the
way we view the origin of life, the water we drink, the death of the
dinosaurs, the creation of the oceans, the nature of the cosmos, and
the very future of the earth itself / by Louis A. Frank with Patrick
Huyghe.
        p. cm.
    "A Birch Lane Press book."
    Includes bibliographical references and index.
    ISBN 1-55972-033-6 : $21.95
    1. Earth.   2. Comets.   3. Life—Origin.   I. Huyghe, Patrick.
II. Title
QB631.F68   1990   90-42763
550—dc20   CIP

Carol Publishing Group books are available at special discounts for
bulk purchases, for sales promotions, fund raising, or educational
use. Special editions can also be created to specifications.
For details contact: Special Sales Department, Carol Publishing
Group. 120 Enterprise Avenue, Secaucus, NJ 07094.

From a drop of water a logician could infer the possibility of an Atlantic or a Niagara without having seen or heard of one or the other.

—Arthur Conan Doyle

Contents

CONTENTS

The Big Splash

Chapter 1

A Radical Departure

I live on the plains of Iowa, drive a pick-up truck, and build instruments for satellites that study the Earth. I am conservative in the way I dress, in the way I lead my life, in the people I vote for, and in the way I do science. Science is my life.

But four years ago pictures relayed from an orbiting satellite led me to a radical departure from the scientific mainstream. I proposed that the Earth was being showered by a vast number of previously undetected, small comets. I calculated that about twenty of these comets, each about the size of a small house and made essentially of water and ice, plunge through our atmosphere each and every minute of every day. If I am correct, then the water that we fish in, bathe in, drink, and are made of, is of extraterrestrial origin.

I did not make this up. The discovery of the small comets is based on a set of observations made at the limits of present-day technology. It so stretches our understanding of the world and impinges on so many areas of scientific study that the interest and ire of nearly the entire scientific community has been aroused. This new population of objects in the solar system has ignited a fierce scientific dispute. If I am right about the existence of the small comets, a whole generation of scientists in a dozen different fields will have been working from geophysical and astronomical assumptions that are simply not correct. The textbooks in a dozen sciences will have to be rewritten.

The proposition that a rain of cosmic ice is presently bombarding the Earth holds staggering implications for us, our planet, and the rest of the solar system. It would mean that our lakes, rivers, and oceans were not formed as we thought early in the Earth's history. It would mean that the substances necessary for the origin of life on this planet may well have arrived from space. It would mean that periodic increases in cometary showers could have caused the Ice Ages and been responsible for the death of many species, including the dinosaurs. It would mean that the planet Venus never had a primitive ocean, as many scientists suspect, and that the Martian landscape was formed by water, which may one day again flow on its surface. And it would mean that the volume of water on the Earth is slowly increasing and may one day submerge the planet entirely. It would mean, in sum, that our cherished notion regarding the Earth's isolation from the rest of the solar system and the universe will have to be discarded.

We once thought the world was small and flat. Columbus sailed across the Atlantic and enlarged it, Magellan sailed around the world and made it round. We thought that the world was at the center of the universe until Copernicus restored the Sun to its proper place and set the Earth in orbit around it. We thought the Earth and its Moon were unique until Galileo showed us that there were similar planets and moons elsewhere in the heavens. And we thought, until the turn of the nineteenth century, that not even a stone could fall from the sky, that nothing could shatter our isolation from the cosmos.

We now accept that the world is round, that it is one of many bodies that circle the Sun, and that, on occasion, a stone does fall from the sky. But science and history continue to maintain that the Earth is safe and secure from the turmoil of the cosmos. We may have bad winters, volcanic activity, and man-made catastrophes, but these are terrestrial affairs pure and simple. The cosmos has no effect on us. We are different. We are isolated.

Life is full of unwritten rules and this is one that science holds dearest. Any tampering with the idea that we are isolated, any suggestion that places the Earth in a relationship with the solar system and universe that is different from the one we have always been taught to accept, will be strongly condemned. We are comfortable with the illusion that the Earth is isolated both in space and time. So if you propose—as I have—that something from out there is affecting us here and now, rather than millions of years in the past or millions of years in the future, beware the wrath of orthodox science.

Science guards our isolation well. Talk about a rock fifty miles wide that resembles a chocolate-covered cherry and lies at the other end of the solar system and scientists will say, "Oh, how fascinating." But talk about bodies close to the Earth and the reaction will be quite different. Say that there are dark objects falling from the sky, which we have never had the capability to see before, and that over the next ten-thousand years these objects will add another inch of water to our planet and scientists will object quite strenuously. Their safety will have been attacked. "Oh, no," they will say, "the water has been on the Earth since the beginning. Earth is here as it is."

Our isolation from the cosmos is staunchly maintained by the guardians of our culture. In 1985, for instance, an editorial appeared in *The New York Times* reflecting on a debate taking place among scientists about whether or not a distant companion star to the Sun could have set off a chain of events that led to the extinction of the dinosaurs. "Astronomers should leave to astrologers," the editorial said, "the task of seeking the cause of earthly events in the stars." The newspaper of record could only express ridicule at the notion that actions in the cosmos could have such a profound effect on the Earth, even though the event in question was said to have happened millions of years ago, far enough removed in time, you would think, to avoid any strenuous opposition. But such things are not possible, the newspaper editors were saying, because the Earth is isolated.

I, too, was taught that planet Earth spins its way through the void, essentially untouched by the chaos of the cosmos. But we know so very little about our solar system, about the interplanetary space that the Earth moves through, about the origins of its water, its carbon, or just about anything else. From elementary school on we read textbooks that make definitive statements, but we just do not know. These are just impressions we have. But I want to know. I would like to understand why we are here. What in the Earth's history, in the history of the solar system, has made it possible for us to be here? It is marvelous the way we inquire about our environment. It is wonderful how we are able to perceive and reflect on it. But just how do we perceive it, and do we perceive it correctly?

If you look at the Earth from space with the eyes that we put on satellites, you can see that this planet holds a unique place in the solar system. Even discounting the fact that there is life on it, it is the most exciting planet in terms of its composition and position relative to the Sun. As occupants of this planet, we should probe further into what makes it so unique, how we got here, and what our relationship with it is. But we should awaken from the rosy dream that we exist separate from our past and future, from the solar system and the rest of the universe.

I think this is the final step in our education about the Earth. This is the next great debate. I do not believe the Earth is a fixed body, unchanged over billions of years. Changes are taking place continuously. The creation of the planets and other celestial bodies is not something that stopped billions of years ago. It is not something that only happens elsewhere in the universe. It is happening here, today. And the small comets are very likely a part of this process.

The comets I have found are much smaller, darker, and far more numerous than the large comets that are known to be wandering into our solar system. Their impact on our perceptions will also be infinitely greater. Yet for some reason it seems never to have occurred to anybody that there might be small comets. We know there are large rocks, large meteors that enter the Earth's atmosphere, as well as small ones. Everyone accepts that now. So if there are large comets, why not small ones? But there is next to nothing in the literature about small comets. The topic essentially does not exist. Until recently the capability to observe such objects has not been available, so without observations to establish their existence, the possibility of small comets in the solar system has been ignored.

In contrast, the known, large comets have been exhaustively studied. They are believed to be primitive remnants of the interstellar gas and dust that condensed to form the solar system. Their nuclei measure about a mile across and have the surface area of a small city. Warmed by the heat as they approach the Sun, their surfaces vaporize and release long tresses of gas and dust known as comas. These comas extend for millions of miles from the cometary nuclei. The word comet, in fact, comes from the Greek word meaning hair. But not all known comets appear to us with these long flowing tresses. Most comets are no more than bright smudges in the sky and are only visible through telescopes. Less than a thousand different large comets have been recorded in all of human history.

You would think that astronomers looking for fainter and fainter objects would have found increasingly smaller comets, but this is not the case. Most astronomers suspect there is a real lack of comets smaller than a football field. They do not know why. They just say so. Small comets, they believe, would most likely be fragments produced by the splitting and breaking-up of large comets and would vaporize and dissipate very quickly. Such brief lifetimes, among other factors, would make small comets very difficult, if not impossible, to find. Unlike the large comets which are detected by their comas, small comets would only be detected by reflected sunlight. And because they are so faint, they would have to pass much closer to the Earth than the large comets in order to be seen. But if they do exist, a few astute astronomers have realized, there would certainly be many of them, and very likely they would be indistinguishable from the meteors we see entering our atmosphere.

This view is essentially correct, though no one would ever have thought that these objects would be as small or as numerous as I have proposed. As a consequence the discovery of the small comets has turned into a broad scale intellectual battle among scientists at the forefront of science. There have been many brisk exchanges in public, but also much intrigue, backbiting, and small-mindedness behind the scenes. What I have experienced in trying to bring this discovery to light can best be described as a game of hardball science. Fairness has had nothing to do with it. Because losing has meant losing big, everyone has played to win.

So there are really two stories in this book. One is about the discovery of a new population of objects in the solar system. The other concerns how science works today, and in particular how controversial ideas are handled by the scientific community. Science is a mountain from which we can see far and wide into the surrounding terrain, the German physicist Max Planck once said, but the mountain itself, it is invisible. I hope that this story will shed some light on that mountain. It is important for all of us to understand just how science works, as it is one of the most influential institutions in our culture. Science shapes the way most of us think about the world.

The past four years have been a difficult time for me, but the truth, as I see it, is this: The infall of small comets into the Earth's atmosphere is taking place today. It is not an isolated event that took place billions of years ago or something that will take place billions of years in the future. It is not something occurring far away on another planet. This is a story about our perceptions of what is happening here and now, what it might have meant for our past, and what it might mean for our future. It has to do with the Earth. It has to do with the oceans, the water you drink, the atmosphere you breathe. It has to do with you. It is a very emotional topic.

Chapter 2

Cosmic Rain

Every year the Great Fool sets his trap. Without fail, one of the news stories of the day is a prank. No one expected this year to be any different. It was the 1st of April, 1986.

On this day France removed its observer troops from Lebanon. The Central Intelligence Agency admitted to have long overestimated the yield of large Soviet nuclear tests. Wildfires continued to blaze across the southeastern portion of the United States. Navy salvage divers sixteen miles from Cape Canaveral continued to haul up more silt-covered wreckage from the space shuttle Challenger. And stock market analysts called the day's twenty-eight point drop a "plunge."

One story bore the mark. It read like something from a B-grade science fiction thriller. A scientist claimed to have evidence that the Earth was being pelted by twenty, hundred-ton comets per minute. These comets have been dumping water on us for more than four billion years, said the scientist, and in one swoop of mathematical magic, he pointed to the oceans and said "Voila!" I am that person. The English tabloids called me all day asking if this was an April Fool's joke. I told them it was no joke. But it did not seem to matter. The Fool had taken my story and made it his own.

I do not really know why the story broke on that date. I suspect that newspaper editors ran the story on April Fool's because it "fit." We had written up our findings in a pair of papers that would be published in the April issue of a refereed scientific journal called *Geophysical Research Letters*. Alex Dessler, the editor of the journal at the time, had sent early copies of our papers to the media at the end of March. Dessler was looking for publicity and he got it. I hate publicity. I do not like reporters and photographers interrupting my isolation and my work. But Dessler thrives on publicity. I think Dessler got more publicity out of this story than for anything else in his life. I certainly have.

I was accused of making a mountain out of a molehill. Critics gleefully quoted Mark Twain's famous remark about science: "One gets such wholesale returns of conjecture out of such a trifling investment of fact." But what else could people say about someone who claimed to have discovered the existence of celestial objects never before seen by astronomers, yet which may be responsible for the origin of the oceans, the origin of life, the death of the dinosaurs, the unexplained falls of ice from the sky, as well as the water in Venus's atmosphere, the mysterious bright flashes sometimes seen on the Moon, the channel-marked surface of Mars, the ice moons of the outer planets, and a host of other phenomena throughout the solar system. My evidence, the so-called molehill, was a bunch of dark spots that had appeared on some satellite pictures of the Earth. It was, I admit, all quite hard to believe.

And very few people did. "That's as crazy as they come," said one astronomer. "I'm amazed," said an oceanographer, "I've never heard anything like it." One geologist dismissed it as simply a case of Halley's fever. But the long-awaited and much publicized flyby of Halley's Comet that year fizzled and my proposed comets had nothing to do with that celebrated non-event. My comets are not occasional visitors. They do not come by every seventy-six years without dropping in. No, these small comet-like objects slam into the Earth's atmosphere at the rate of ten million a year.

The evidence that led us to propose such a massive infall of objects into the Earth's atmosphere came from a high-altitude, polar-orbiting satellite called Dynamics Explorer. The satellite carries an ultraviolet camera which takes pictures of the Earth in the light which lies just beyond the violet end of the visible light range. These otherwise normal ultraviolet images were speckled with some very puzzling dark spots. We found more than 30,000 such spots in the images during some 2,000 hours of observing time.

At first I thought the spots might be due to faulty instrumentation. So with the help of John Sigwarth, one of my graduate students, and my colleague John Craven, a research scientist at the University of Iowa, we scrutinized the data for possible computer glitches, random flaws, radio transmission noise, failing sensors, even the possibility of paint flecks on the camera. But after years of checking and rechecking, testing and retesting, and having eliminated the possibility of spurious effects, we had to face the fact that the dark spots were real.

It was clear that the spots on the images represented something that lay between the satellite itself and the region of the upper atmosphere we were observing. Something was absorbing ultraviolet light in that region and creating a kind of dark "hole" in the uniform glow of the Earth. The signature, by all appearances, indicated the presence of water. But these dark spots were enormous, measuring about thirty miles in diameter. No amount of evaporation from the surface of the Earth could produce such large clouds of water vapor at that altitude. The water had to come from *out there*. The source clearly had to be extraterrestrial.

But I knew that a cloud of water vapor in the interplanetary medium would disappear very quickly. So whatever had caused this large cloud in our images had to be in a form other than water vapor when it was well away from the Earth. I calculated it to be a relatively loosely-packed ball of water-snow the size of a small house. At one to two thousand miles above the Earth either the action of tidal forces, which are produced by the planet's gravity, or perhaps static electricity, would then break up this object and the Sun would vaporize its contents into a ball of gas. By the time it had plunged to within about 300 miles of the Earth, this hundred-ton spray of water and snow, traveling at about twenty times the speed of sound, would have expanded to a thin ball of gas some thirty miles across. That is not much water. It is thinner than a London fog. But it would be enough to produce the black spots in our images.

At about fifty miles above us in the atmosphere, the vapor cloud would slow to subsonic speeds, but continue to plummet until it reached a height of about thirty-five miles where it would get mixed up with the air in the upper atmosphere. Winds circulating into the stratosphere would then turn the water vapor into ice crystals. These ice crystals would fall and become part of the water vapor present at low altitudes, eventually turning into precipitation and falling on land and sea.

This, of course, is the scenario for just one comet-like object. At the observed rate of twenty per minute, over a year's time you would get one ten-thousandths of an inch of water on the Earth's surface. That is but a tiny fraction of the annual rainfall and quite indistinguishable from it. It is a mere drop in the bucket. But the icing on the cometary cake, so to speak, is that if you add up all the drops accumulated over the age of the Earth, you would have enough water to fill the oceans and more.

That is how we came to the startling conclusion that the dark spots in the images were being formed by small, comet-like objects entering the Earth's atmosphere. The mind rebels. It is disturbing. But it is really quite reasonable. I can explain why the black spots in the satellite images cannot be anything else. I can explain why we are not seeing these comets all the time. I can explain just how the water from these comets manages to survive through the atmosphere. And I would have to do just that in the months and years to follow.

The first time I did so in public was in Baltimore a little more than a month after the publication of our two original papers. The occasion was the annual spring meeting of the American Geophysical Union. The AGU, as it is known, has 22,000 members worldwide "dedicated to the study of the Earth and its environment in space." My proposal had sparked a great debate among the community of physical scientists and many were here to apply the pressure, to crush this bothersome fly with a cinderblock if need be, before it created any further damage to the established notions of science.

The meeting room was packed. The seventy or so seats were filled and people were lined along the back wall. A television monitor was even provided so that the dozens outside the room could also witness the proceedings. It was hot in the room and rivers of sweat rolled down my face and soaked my white shirt. I removed my jacket and loosened my red tie. As I walked to the front of the room one of my colleagues told a reporter: "There's nothing like a good, bloody brawl in science."

It was clear why everyone was after me. I had committed two deadly sins. The first sin can only be made by an establishment scientist: I broke from the mold. I was working along with everyone else then suddenly seemed to reverse course. I did something that drastically disturbed what everyone else was doing and thinking. And while my proposal ran over everyone else's work—from planetary astronomy to oceanography—it had no real effect on my own work. That was my second deadly sin. It upset people greatly that this finding had so much of an effect on their work and none at all on mine. But I was caught between the proverbial rock and a hard place. People said I should not be wandering over all these fields. That was true. But I had no choice. I had to defend myself.

I could understand the reaction of my critics. Many of them, by the way, happen to be my friends. I would not have reacted any differently. A lot of people said it was impossible. That did not bother me. The existence of such a large number of small comets is difficult to grasp. It introduces a large, previously unknown factor, at least from an observational point of view, so it is upsetting for people in many fields. If it is correct it is going to change an awful lot of things. Its implications are widespread.

I am, as I have said, a very conservative person. One of the things that bothered people so much about my proposal was that this was just totally out-of-character for me. Of course, they did not consider how many years of effort I had spent on it. I knew this was a hot potato, and I did everything I could to get rid of it. Many people thought my career would be ruined. "What if you're wrong?" they said. But being wrong is probably not as great a risk for a scientist who has already established his reputation. I was not going to be destroyed by it. If this work had been done carelessly, certainly my career would have been ruined. But most people would agree that I have not been careless about this at all.

I have no complaints about the harsh treatment I have received. I expected it. Many of the arguments were clever and on the mark. But unfortunately much of it had nothing to do with the science at all but with belief. I harbor no bad feeling about it. I asked myself two questions before I published the papers. Was I old enough and did I know enough to be able to carry out this debate? I was forty-seven at the time, and I decided that I was old enough and knew enough, however little that was, to be able to do it. But I also wondered if I were physically able to go through with it. I was overweight at the time. I wondered if this debate would send me to my grave.

So I cut down on the weight and proceeded to work continuously for two and a half years. I slept four or five hours a night for months at a time. I took no weekends off, no vacations, no holidays. I developed a hernia at the time, but I had to postpone the operation for several months. I worked until two and three in the morning every day, Christmas and New Year's included. I thought that it was my responsibility. If I was going to cause this much fuss, this much apparent harm to science, then I must pay the price.

Nor did I wish to stop my work in plasma physics—the clumps of charged particles in the cosmos—during the four years that this controversy raged on over the small comets. I did not want to drop from the circuit, the endless rounds of meetings and seminars in my field which, like circuits for golf or tennis professionals, require that players be on the road for much of the year. The circuit is where the newest ideas and data are discussed long before they are published in the journals. If you just publish papers, but never travel, never give talks, and never attend meetings, you cannot be a major player in modern science. It is not a glamorous existence. You are on the road so much, living out of a suitcase, that you need your appointment book to know whether you are in Tokyo or Stockholm. And when you return, you are overwhelmed with paperwork, writing grant proposals to finance your work, and reviewing papers for scientific journals.

It has been an exhausting four years. But one of the things that made this effort possible is that I am isolated here at the University of Iowa. There is no name on my door, so there are no distractions. I do not have a heavy teaching load. The University of Iowa is not a well-known university. It is an intellectual outpost in a soybean field.

Chapter 3

The Black Spot Mystery

The black spots on the images were like flies walking across a television set. They were annoying. They were there from the start, on the very first images. They are still there today. There is no question about who saw them first. Everybody saw them. We would give talks and the black spots were there on the images for everyone to see.

It all began back in 1981. On the third day of August a Delta rocket lifted off from Vandenberg Air Force Base carrying a pair of NASA satellites, both known as Dynamics Explorer, into elliptical orbits. One of these satellites circles the poles of the Earth at an altitude ranging from 350 miles to 14,500 miles. I was responsible for three of the instruments aboard the satellite, one of which is an ultraviolet camera, built and operated by my colleague John Craven.

This satellite was designed to examine the Earth for certain light emissions that are invisible to the naked eye. I and other scientists hoped that these emissions would provide further insight into the nature of the auroral lights that occur in the Earth's polar atmosphere and detect any effects associated with them. I was particularly interested in getting the first global pictures of the Earth's aurora and I was not disappointed.

The pictures sent back from the ultraviolet camera on the satellite were spectacular. The remarkable auroral crowns encircle the poles of the Earth, while the planet's dayside looks like a bright ball illuminated by a flashlight. This bright feature is known as the dayglow. Dayglow is produced by the interaction of sunlight with the atomic oxygen present in the Earth's upper atmosphere. The ultraviolet light emitted by this dayglow is not visible to the naked eye but is within the range of the satellite's specially-designed camera. The emissions it captures are transformed into a normal photograph.

But the images of the Earth we obtained beginning in late 1981 contained an unexpected feature. The blanket of dayglow was not uniform. It was speckled with dark spots. Strictly speaking, these spots were areas of greatly reduced brightness. In other words, there seemed to be holes in the dayglow. Everyone saw them. Everyone assumed they were noise—those random fluctuations in data that are due to chance.

Life went on. During the summer of 1982, an undergraduate student named John Sigwarth began working for me. Sigwarth had come to the University of Iowa in 1979 and had taken his first physics class with me. He was very interested in space science and was very bright. I wanted him to take enlargements of the satellite images and scan them for signs of gravity waves, small scale ripples in the upper atmosphere that sometimes follow the brightening of auroras. These waves sweep across the face of the Earth much like waves in the ocean. They have been detected on radar and I wanted to know whether these waves would show up as crests and valleys of light in the satellite images. I assigned to Sigwarth the task of processing the data to make the waves visible.

But he could not do it. He worked on the computer program, which was designed to extract subtle features from the original pictures and highlight them, but each time the computer stuttered and burped as it scanned those black holes in the atmosphere. The black spots kept getting in the way. Then one day in the fall of 1982 Sigwarth, quite exasperated, walked into my office on the second floor of Van Allen Hall and said: "How can we get rid of this stuff?"

Sigwarth, Craven, and I pondered the question and concluded there must be some disturbance in the camera's electronics that would occasionally put these annoying little black spots in the image. Perhaps a transistor was fluky, or a computer was not functioning properly, or a problem arose as the pictures were being transmitted down to us from the satellite. It was possible.

Each picture involves the transfer of an enormous amount of information. Like the image produced by a FAX machine or by the display on a home computer, each of the satellite pictures is constructed of a large number of little dots called pixels. Each of these pixels is described by eight "bits," actually a string of "0's" and "1's" that indicate the intensity or brightness of that pixel. So perhaps some of the image pixels were not being transmitted down properly from the spacecraft. Whatever the source of the noise, I did not relish spending my time tracking it down. Sigwarth got the job.

It was tempting to simply remove the spots from the images and get on with the search for gravity waves. But you cannot alter data on a mere assumption. You have to have a reason. We needed to show that the spots were either detector noise, or produced by electronics on the spacecraft, or generated by computers on the ground. Only once that was accomplished could we eliminate the spots from the processed images and get on with our work.

Sigwarth worked very hard trying to solve the mystery. From time to time he would come into my office and say he was not having much luck with it. The year 1982 drew to a close. Sigwarth kept working on it. But he was unable to trace how the holes appeared on the images. One possibility that we entertained was that one of the two light counters on the camera was failing. Every other pixel in the satellite images is produced by an entirely separate set of electronics. The counters, in other words, take turns producing the dots that comprise each image. So Sigwarth had separated the data produced by each counter. But when he examined the data he saw that both counters were observing spots in the same sequence and at the same rate. This told us the counters were not dropping bits, those meaningful strings of "0's" and "1's." We knew that it was nearly impossible for two counters to malfunction in exactly the same way.

We also eliminated the possibility that these annoying little black spots were caused by errors in radio transmission from the spacecraft. We checked the entire system from the time the data left the instrument, passed through the satellite itself, traveled down to the ground, and was relayed to us. Because the instrument regularly transmits fixed words, or fixed bit patterns, we could check to see if any transmission errors were occurring. We calculated that dark spots due to telemetry noise would appear once in every 200 images. But these spots appeared in the images almost a thousand times more often than expected.

All of the tests we performed on the data produced results that ran counter to our expectations. We really wanted to show that the holes were noise. But we were out of luck. So Sigwarth began to look at the pictures in great detail, trying to see if a spot that appeared in one frame could be seen in any of the subsequent frames. A very careful analysis showed that the spots could be followed in this way, although the spots in the subsequent exposures were not as dark as those in the initial exposures. This seemed to indicate that the black spots were moving and changing.

Sigwarth then programmed the camera so that instead of scanning the entire Earth, it would scan just a small portion of it. This allowed the camera to return to the same area more quickly. The series of pictures produced showed that a black spot would appear and disappear in a sequence of frames. The black spots seemed to be objects in motion. This was not characteristic of noise. Noise should appear at random all over the image. This indicated the presence of a real object. Sigwarth came down to the lab where I was working and showed me the data. He was very excited. I looked down at the pictures and congratulated him. I thought we were on to something.

Other clues convinced us that the spots were genuine. The spots, or holes in the atmosphere, appeared to move in the same direction across the face of the Earth. If these holes were random events, due to malfunctioning equipment, for instance, you would expect to find half the spots going in one direction, and half going in the other direction. But this was not the case. Most of the holes appeared to move in the same direction across the face of the Earth.

By February of 1983 we had come to the conclusion that something, some kind of object, was absorbing the ultraviolet radiation between the camera on the satellite and the Earth and producing the apparent holes in the atmosphere. The more we looked the more it seemed that our images were actually snapshots of the movement of these objects above the atmosphere. We began to suspect that these objects were meteors of some unusual sort.

So we decided to compare the motion of our elusive black holes to the passage of meteoric dust and debris in our atmosphere. Much of this meteoric dust tends to orbit the Sun more rapidly than the Earth. It essentially catches up to the Earth, in other words, and approaches the atmosphere from the local evening face of the planet. Such motion relative to the Earth is called prograde motion. Whatever the black spots represented, they showed the prograde motion that is characteristic of meteoric material. This not only implied that the spots were real but that the objects they represented were extraterrestrial. So we assumed, given the large number of spots that showed up in our images, that the holes were caused by a particularly large influx of meteors such as you would have with a meteor shower.

We went public with these results for the first time in May of 1983. Two weeks after receiving his undergraduate degree in physics, Sigwarth presented our findings in a paper entitled "Atmospheric Holes Possibly Associated with Meteors" at the spring meeting of the American Geophysical Union in Baltimore. I sponsored the paper and John Craven was listed as a co-author. People seemed interested and curious, nothing more. But over the next two-and-a-half years we presented three more papers on the topic at meetings of the American Geophysical Union and each time the audience grew.

Over this period of time Sigwarth and I analyzed over 10,000 images and learned a good deal about the black spots in the process. Our interpretation of the events continued to involve meteor impacts into the Earth's upper atmosphere. By counting the spots in our images we were able to estimate the rate at which these objects appeared. This was the simplest measurement to do. We saw ten holes per minute on the daylight side of the Earth. So we doubled that figure to obtain the rate of these objects over the entire face of the Earth. There had to be about twenty such objects entering the atmosphere every minute. That was an alarming number of objects.

We still needed to explain just how these objects, which we assumed to be meteors, could cause holes in the atmosphere's screen of atomic oxygen. We entertained three possibilities. The first and simplest explanation had the meteors laying a blanket of material over the atmosphere, preventing the light from getting through, and creating a black spot in our images. Another possibility was that the atomic oxygen up there was being depleted by some special chemical process. But we could think of no chemical reaction that could get rid of the atomic oxygen quickly enough and none that allowed the atmosphere to restore itself as rapidly as we observed in our images. A third possibility involved a catalytic reaction of the sort that takes place in the catalytic converter in your car. Could some small amount of material, some catalyst, be converting the atomic oxygen in the atmosphere into molecular oxygen and producing the dark spots in our images? It was not likely, as we were never able to identify any such catalytic agent.

The knowledge that the spots actually moved across the face of the Earth strongly pointed to the existence of some kind of an object that prevented light from passing through it. Whatever it was had to be big and blackening out the ultraviolet light at a certain wavelength. It could not be an atom. It could not be a rock. It could not be anything thrown up there from down here. It had to be a common molecule in the solar system that absorbs at the right wavelength. The only common molecule is water and water just happens to absorb at the wavelengths we were observing with our camera. There was no reason to look for anything exotic. Water, in the form of water vapor, fit the bill perfectly.

This explanation posed certain difficulties, all of which were more psychological than physical. When we calculated how much water we would need up there to produce a spot in our images, we came up with a figure of about a hundred tons. Anyone would tend to back off from such a large figure and initially we did too. Then we figured out how many such objects we needed to account for the holes in the images we observed over the course of the year. And it was not one, not a hundred, but ten million. There was the problem. One per year would not have been a problem. But ten million per year? Unfortunately, there was not much leeway in our numbers.

The size of the holes presented another problem. They were easy enough to measure. We knew the size of the area each pixel covered in our pictures and we knew the altitude of the space-craft. But what looked like little dark spots on the images turned out, in reality, to be about thirty miles across. They could not be rocks because such large rocks would just smash the surface of the Earth to pieces. These were clouds of water vapor.

It was the only reasonable explanation we could find. It was the only reasonable explanation that anybody has ever been able to offer. Early on some engineers suggested that the black spots might be due to satellite parts falling into the atmosphere. But that is not possible. First of all there are not ten million satellite parts out there falling down on us each year. And secondly, they are not thirty miles across. Many people said there had to be other explanations for the dark spots. But no one has ever come out with one.

The numbers were shocking. Earlier we had found that the spots on our images varied with the frequency of meteors falling into the Earth. At first, we were elated. But when we sat down to think about it, we saw a disaster pending. The objects were just too big and too numerous to be just another nice little geophysical fact with no real impact on our thinking. But once we ruled out noise and ascertained that the spots were real, the next step, the interpretation, was trivial. The only way to interpret these events was in terms of ten million objects falling into the Earth's atmosphere every year. That is an infall of material that is about ten thousand times more than anyone had ever imagined. So psychologically, emotionally, the interpretation was difficult to accept. But intellectually, it was trivial. There was no other reasonable explanation.

By December of 1985 the press had picked up the story. Stephie Weisburd had written an article for Science News entitled "Atmospheric Footprints of Icy Meteors" following our fourth presentation on the topic, which took place at the fall meeting of the American Geophysical Union in San Francisco. We still called the objects meteors at that time, but that was not quite right. One thinks of a meteor as dust and rock, not water, although some meteors are thought to contain a substantial amount of water before hitting the atmosphere. But the objects we were talking about clearly had to be mostly water. That could only mean one thing. Sigwarth and I finally realized after the meeting that these objects could be nothing other than comets, small comets.

People at the meeting seemed to be quite interested in our presentation but we were careful not to threaten anyone's perception of the solar system. We never came out and said just how big these objects were or exactly how many were falling into the Earth's atmosphere. We just said there were a large number of unknown objects falling into the Earth's atmosphere that had not been detected before. I also began discussing our findings privately, in other meetings and hallways. No one got too upset over it.

Science is no stranger to odd data. Sometimes anomalous observations are just put aside for later examination. But often they are just stuffed into a drawer and forgotten. We scientists normally have our hands full examining the data we were seeking in the first place. So odd data has a tendency to fall through the cracks. There is a lot of tucking under in science. Everyone thought I would eventually bury my data on the mysterious black spots. No one thought I would publish them.

They were wrong. I began writing the first draft of the original small comet papers in December of 1985. Eventually I handed out a few copies to people for their comments, but not before spending many nights pacing the hallways of Van Allen Hall and pondering the consequences of such an action. The papers touched on many of the objections that such a large infall of comets would raise. What would their existence mean for the Earth and Venus and Mars and the other planets? What was the lifetime of such objects? What kind of mantle did they need to survive repeated passages through the inner solar system? Where did they come from?

One of the copies of our original papers had gone to Thomas Donahue, a professor of atmospheric sciences at the University of Michigan. You cannot do any better than Donahue when it comes to experts on planetary atmospheres. I respected his opinion. Donahue looked at our papers and said that if we were right these comets would have brought in enough water, over the age of the Earth, to produce the oceans. And of course he was right. I had not thought of it because I had always assumed that the oceans had always simply been here. So did everyone else.

I submitted my first papers on the small comets to Alex Dessler in February of 1986. Dessler had just become editor of the American Geophysical Union's *Geophysical Research Letters.* He was ambitious and wanted to make the journal controversial, exciting and well-known. Unlike the quarterly *Reviews of Geophysics,* which Dessler had edited previously, this monthly publication provided a rapid way to communicate the latest research in the field of geophysics. Shortly after seeing our papers Dessler called to say that this was just the kind of material he wanted. He would have the papers reviewed in forty-eight hours.

Scientific journals rely on a rigorous review process to weed out inaccurate claims and research findings. Editors, if they are favorably disposed to manuscripts submitted for publication, send them to two or more scientists considered experts on the topic treated in the manuscript. The comments of these referees are returned to the editor, who then sends them on anonymously to the original author. This allows the referees to be candid and honest. Often, however, this power is abused. Those who are judging the merits of a manuscript are often in competition with the author for grants or recognition. Some scientists abhor this anonymity and those who know one another occasionally send their comments directly to the author of the submitted paper.

I had this kind of unspoken understanding with Donahue. He was one of the referees and he nearly exploded. He asked me not to publish the papers, the interpretation paper, in particular. The first paper was simply a description of the black spots and no one could deny that the black spots were there. The second paper was the interpretation. In it we spelled out that we were dealing with ten million, comet-like objects entering the Earth's atmosphere per year, each one the size and weight of a small house. We also touched upon many of the topics this interpretation would seem to contradict, such as the origin of the oceans, as well as the lack of water in the Earth's upper atmosphere, on Mars, on Venus and on the Moon, as much as we could cram, in other words, into four pages. This was the limit on the length of papers published in *Geophysical Research Letters.*

The other referee also recommended against publication. He said, Dessler told me, that "if this was correct, we would have to burn half the contents of the libraries in the physical sciences." It was a dicey situation for Dessler. The two people he had asked to review our papers had advised him not to publish. So Dessler called me and asked if I would withdraw the interpretation. He warned that its publication might destroy my scientific career. But I told him it was necessary. Without it people were likely to regard the black spots as a mere curiosity.

Donahue also begged me to come to my senses. He did not want the interpretation paper published. I think he was trying to be helpful. But he must also have been worried about the tremendous repercussions these findings would have for many fields of science, including planetary atmospheres, Donahue's own little niche. Yet despite two negative reviews and the "uncomfortable ramifications for the community at large," Dessler decided to go ahead with publication. The final decision always rests with the editor. Reviews are designed to guide, not bind.

"If you restrict the journal to publishing only what pleases the referees you end up publishing what is popular," Dessler explained in 1987. "And while it does make everyone feel more comfortable, you are guaranteed to miss the occasional breakthrough. There is a price to pay, of course, for doing it the way I think it should be done. Nine out of ten of these things will turn out to be rather worthless. But I think the price is worth it. Occasionally one is so important that it makes publishing all the previous ideas that didn't turn out quite right worthwhile."

Anytime you break new ground in science you get attacked. This is the basic conservatism of science. But it is necessary to put all new ideas on trial. I had some notion of what the response to these papers would be. I had been involved in many scientific debates before. But the reaction I received was like none I ever experienced. I was driving a bulldozer through dozens of the neatly planted fields of science and everyone was upset.

I had, of course, been tempted not to publish these results. I was enjoying my work. People respected me. I had all the opportunities in the world and all the funding I needed. I did not need the aggravation these findings would stir up. On the other hand, I thought of all the scientists from Copernicus to Einstein who had come forward with seemingly outlandish but revolutionary results. I realized that if I was right and did not publish my results—and no one did this work for another ten or twenty years—then I would be wasting thousands of man-years of scientific effort.

People had warned me repeatedly and I understood what they were saying. Had I heard someone talking the way I did, I would have warned them as well. But I had agonized over the decision to publish this material for a long time. The decision came long after it became quite apparent that our efforts to show that the black spots were not real had failed. There were no other tests to be done. I knew this would have a big impact on science. But I felt I had no choice. I could not bury this material.

The notion of so much water falling in from space flew against all current observations and beliefs. Most likely the notion was wrong, so I had to make certain that it was not. I had done everything I possibly could to find something that said these objects could not exist. I only needed one big piece of evidence. Just one. A hundred pieces of evidence would not prove that they exist. But it would only take one to show that they did not. So I sat in libraries and read about astronomy, about oceanography, about geology, one field after another. And finally I decided that our findings must be published, no matter what the con-sequences. I could not live with myself otherwise. It was just morally incorrect.

Chapter 4

True Confessions

I was born in Chicago in 1938, the year "You Must Have Been A Beautiful Baby" became a popular song. We lived in Berwyn, Illinois, until I reached first grade. But my parents were concerned about the violence around Chicago. They thought that a better place to raise children might be somewhere out on the plains, so we moved to Topeka, Kansas. We lived there with my grandparents for a short while and then moved to Chanute, Kansas. I went to grade school there.

I was the oldest child. I have a brother, who is younger by two years, and a sister, younger by six years. My father worked for the Santa Fe railroad. We did not have much money, but I never knew we were poor. My father never made it through high school because he had to support the family. At an early age I was trained to be a pianist. I took piano lessons and played in a very emotional way.

School bored me. I sat in the classroom and seldom paid any attention to what was going on. Teachers ostracized me. Whenever we had spelling bees or math contests I competed against the last person and won. So the students ostracized me as well.

Just before junior high, we moved to Fort Madison, Iowa, a rivertown on the Mississippi. I spent my time reading science fiction and building things like cannons. I also continued to play the piano. By the time I got to high school, I was known as a problem child. Much of the problem was my poor memory for dates and names. Once I was registering for classes and could not fill out the forms. I could not remember the date of my birthday.

The principal sent me home to find out when I was born. He thought I was mentally retarded. When I returned, he had signed me up for non-academic courses, like machine-shop, because he thought I would never make it to college.

High school presented no challenges. The teachers sometimes asked me not to attend classes because I depressed the students by my performance. I actually taught the chemistry and physics classes when the instructor was not there. I got every award you could get in high school, everything from the typing award to the American history award. I would probably also have gotten the home economics award but they would not let me take the course. I was class valedictorian.

My mentor was a chemistry and physics teacher named William Heinberg. He came from industry to see if he could teach. He actually stayed on an extra year, before going back to industry, just so that he could teach me chemistry and physics. He had a profound effect on my life. He trusted me to work in the labs and I became fascinated with science.

I went to college at the cheapest place I could find and that was the University of Iowa. I thought I would never again have the same problems in class. But I was not there a month before my English teacher told me that if I did not come to class, he would give me an "A." He said I discouraged the other students. But I loved science. I changed from chemistry to a physics major because that appealed more to me.

I enjoyed my freshman year. That was 1956. I needed work and found a job in the physics department helping to plot points for someone's doctoral thesis on cosmic rays. I enjoyed reading the data from the printouts and constructing graphs. Then, before the start of my second year, Professor James Van Allen called me into his office and offered me a graduate teaching assistantship. I was a sophomore undergraduate at the time. I was a kid. But I did it. Everyone in the class was older than I was. You had to take ROTC at the time and I would come into class in my ROTC uniform. It made me look even more like a kid. But we all had a really good time.

By 1958 the space program was underway and the University of Iowa was in the thick of it. Van Allen's lab was in the basement of the physics building. Things were stored in dog cages and at one in the morning bats flew around the hallways. It was like no lab you see today. Van Allen was in charge of the Geiger counter that would be placed on Explorer 1, the first U.S. spacecraft. It weighed just over ten pounds and was about the size of two oatmeal boxes. Explorer 1 was launched on the last day of January and discovered the radiation belts that gird the Earth. They now bear Van Allen's name.

In my junior year Van Allen asked me to help him calibrate the instruments for two other spacecraft, Pioneers 3 and 4, which would be launched atop modified ballistic missiles. Pioneer 4 would become the first successful lunar probe. The Pioneers were the size of an oatmeal box and shaped like a cone. Each spacecraft contained a battery, transmitter, and Geiger counters. Our interest in these spacecraft was intense since the day in October 1957 when we heard the news that the Russians had launched something called Sputnik. We had strung an antenna up on top of the physics building and tuned in as Sputnik passed overhead. It made a "beep beep beep" sound that I will never forget. I was impressed. But I was also distressed that the United States would not be the first into space.

The Pioneers were launched during my undergraduate junior year. Pioneer 3 did not have enough "oomph" to make it to the Moon. It only got about half way there but it did make some beautiful measurements of the far boundaries of the radiation zones that had been discovered by Explorer 1. I spent a lot of time plotting and analyzing the data. Van Allen wrote the papers. I was barely twenty years old and I co-authored two scientific papers with him. Van Allen and Frank. Those were very exciting times.

I received my bachelor's degree in 1960, but continued on as a graduate student, spending many hours in the lab. The space sciences fascinated me. I published a number of papers, both on my own and with others. I got my master's and Ph.D. in about four years. I was undecided about what to do next and was interviewed by several places. The day I went for the interview at Berkeley, President Kennedy was shot. But the opportunities at Iowa were so great at that time that I decided to stay. By the age of thirty-two I was a full professor.

I have had experiments on about forty spacecraft. People tell me I have had more instruments on more spacecraft than anyone else on this planet, but I have never bothered to check that out. The point is simply that I love experimental physics. I like to build things and use the best technology available to probe our environment and solar system and expand our knowledge. My area of expertise is plasma physics. Plasmas are the charged particles that comprise 99 percent of the universe. You look up in the sky and the stars are plasma. Our Sun is plasma. The Earth's aurora are caused by plasma. In order to understand the universe, you must understand plasma. That is what I try to do.

I have been in the forefront of many plasma discoveries. I made the first measurements of the plasma ring around Saturn. I was the first to measure solar wind plasmas funneling directly into the Earth's polar atmosphere. I was the first to observe with a satellite instrument the belt of ions around the Earth that is now known as the "ring current." And I discovered the theta aurora, a luminous phenomenon which, seen from space, looks like the Greek letter theta stamped across the polar cap.

If you know enough about experimental physics, you can work in any field you like. That is because physics deals with fundamentals. I have always hoped to be in a class of people known as tinkerers. These are people who see something in a field outside their own that so stimulates their imagination that they make the necessary measurements in the hopes of pressing forward the frontiers of knowledge. That to me is the final stage, the ultimate achievement, of an experimental physicist.

My whole life has been driven by my work. Unfortunately, it has come at the expense of my family. I have not been a particularly good father or husband. I have not spent a lot of time with my two daughters. I am thankful that they have a fine mother who has raised them well. During the past few years, my daughters have seen me more on TV and in the newspapers than in person. They talk to my secretary more than they talk to me. That is not good, but this is the way I am. I realize that this is selfish on my part, but I am just consumed by what I do.

There is a wonderful restlessness to science. It keeps wanting to move on, to explore new territories. Since the beginning of the space age we have moved rapidly to explore the Moon, the planets of the solar system, and the furthest objects at the very edge of the universe. But for most of my career I have looked back at the Earth. I have been particularly fascinated by how nature produces those beautiful little crowns, the auroras, on the north and south poles of the Earth. The Earth and its royal crowns are every bit as magnificent as Jupiter and its big red spot. These space-borne views of the Earth have provided many insights to similar phenomena observed elsewhere in the solar system.

I am presently involved as an experimenter on the Galileo mission to Jupiter. My instrument will measure the plasmas at Jupiter. I am building what is probably the most advanced dim-light imager in the world for the Polar spacecraft to be launched in 1993. I am one of two U.S. experimenters putting entire instruments on a Japanese spacecraft known as Geotail, due to be launched in 1992. I have a lot to do.

People generally do not equate the "L. Frank" who works in plasma physics with the "L. Frank" who discovered the small comets. One appears to be the most conservative of scientists, the other a maverick hellbent on destroying the very foundations of science. But they are one and the same person. The small comets are only one aspect of my life's work. In the past five years, I have managed to author or co-author about sixty papers, only some of which have been on the small comets. I did not want, and could not afford, to let the small comet controversy stop me from my other work.

There is no easy or clearcut blueprint for scientific discovery, but the process, it seems to me, usually follows one of two paths. It can occur, in one case, when you are pressing forward the frontiers of a science. This requires that you devote your life to one very narrow scientific topic. Take solid state physics, for example. The current frontier in that area is superconductivity— the ability to conduct electricity without resistance. If you are working in this field you are trying to find the right combination of metals and alloys that will conduct electricity at higher and higher temperatures. If you are interested in making a supercon-ductor, you are not reading about geology, or astronomy, or even what is happening in high energy physics. You do not have the time. You are working in your little niche day and night, knowing that there are dozens of other people competing with you. There are ten, twenty, thirty other large laboratories, not counting the many universities, trying to do the same thing you are. You are going to feel the pressure, the excitement of it. But although there may be a temporary clash of ideas, a debate perhaps, there is no conflict.

Conflicts occur when your findings clash with accepted scientific dogma. This is the second path to scientific discovery, but it is generally not a pleasant experience. Those who question the prevailing point of view are usually regarded as outsiders, as mavericks, as renegades. Their behavior is often seen as dogmatic and pathological. And the scientific community responds by either turning on you or turning away from you. The history of science is riddled with examples.

Hannes Alfvén, the Swedish astrophysicist now in his eighties, is one of those who was essentially ignored. He developed the fundamental principles of plasma physics in the 1940s but was not recognized for his work until three decades later. He won the Nobel Prize in 1970. But at the beginning, no one paid any attention to him. His work opened up a whole new area of science. Yet his very sophisticated ideas about the aurora, sunspots, solar wind, the flow of plasma, and about the origins of the solar system were far enough removed from scientific thinking of the time not to upset any applecarts. Even today, with such concepts as "Alfvén waves" firmly entrenched in space physics, the man is still regarded with ambivalence. Scientists respect his past contributions but remain skeptical of his theories.

Alfred Wegener suffered the opposite fate. In 1912 he proposed the revolutionary hypothesis that the continents are not fixed but rather have been slowly wandering during the course of the Earth's history. This proved far too radical a notion for most of his contemporaries to accept. The reaction was heated. Everyone turned against him. Wegener, who was occupied most of his professional life as a meteorologist, had no time to respond adequately to the increasing opposition facing his theory of continental drift and was essentially drummed out of the scientific corps. Wegener died in shame because he thought he was wrong. Vindication did not come until many years later. But even today there are still people who do not believe in his theory of continental drift.

I think that I understand how Alfred Wegener felt when he

proposed his theory. He told people that the ground they were standing on was moving under them. That hit close to home. It cut across a lot of areas of science. I am telling people that tens of thousands of small comets are falling into the sky above them each day. That also hits close to home. And it also cuts across dozens of areas of science. No one plows over everybody's neatly planted gardens without paying the consequences.

Chapter 5

A Storm of Controversy

Even before publication of our discovery, the small comet proposal became the target of vigorous and even vociferous attacks. The battle lines were drawn quickly and easily—it was us against everyone else. No blood would be shed, but in the months to come a sea of ink would cover the battleground, the pages of *Geophysical Research Letters*. We had to answer every question that was thrown at us regarding the small comets and we had to do it quickly.

Our discovery drew a flood of critical comments. The editor of the journal, Alex Dessler, had welcomed controversial topics to its pages and had established a specific procedure, resembling the pre-arranged rules of a duel, for handling any formal critical responses to the published articles. The attack of the challenger he called the Comment. Dessler gave it an initial reading, but it was not refereed. Comments judged to be valid were then forwarded to the defender.

Dessler required that the defender meet the opponent's attack with a counterattack, labeled the Reply, but he would allow no ducking or side-stepping of the issues raised by the Comment. This parrying continued—with as much skill, speed, and finesse as the players could muster—until one side or the other chose not to revise its last submission. Dessler then published the final drafts of the Comment and Reply together in a future issue of the journal—point pricks, shoulder jabs and all. Dessler hoped that this would present readers with an accurate summary of the match, "a terse distillation of what might have been many pages of private debate."

For *Geophysical Research Letters*, the interest stirred up by the small comets was unprecedented. The first Comment reached Dessler's desk before the original papers announcing our discovery had even been published. Less than a month later, Dessler had accepted four Comments from the critics and within three months he had accepted ten critical Comments for publication. It would take more than a year out of my life to answer them all, a total of eleven Comments in the end. It was like a paddling line. There was no rest. If I had been ten years older, I would have fallen apart physically. As it was, I barely made it.

Sigwarth and I did the best we could to answer this laundry list of objections in as orderly a fashion as possible. I would receive the Comments, since I was listed as first author, and would make a copy for Sigwarth. He would read it and we would discuss the various points in the Comment. We then decided what we needed to do in order to answer the question it posed. Sigwarth scurried around several libraries on campus, digging out material that often had never been checked out before—or since. We reviewed the material together, learning about the topic raised by the Comment. Sigwarth would do most of the computer work. We both did the theoretical calculations and discussed our results with Craven. Then I wrote the Reply.

The Comments received by *Geophysical Research Letters* defined the scope of the small comet debate. They covered a very broad range of subjects from geology to planetary atmospheres and from lunar science to astronomy. The entire point of this exercise was to find inconsistencies with the small comet hypothesis. If one was found to contradict the existence of small comets, these objects would then cease to exist. If I faltered, if I failed to answer any of the Comments, if I could not come up with a half-way reasonable physical explanation for the criticisms that were flung at us, then it was over. In fact, Dessler called me on occasion and said that I did not have to answer the Comments. But everybody in the world knew that if I failed to answer them, then it was an admission that the debate had come to an end.

Dessler arbitrated the debate. Sometimes several Comments made the same point. Not all Comments were accepted. The author of the first critical Comment to reach Dessler's desk and the first to be published was Thomas Donahue, one of the two referees of our original papers. Donahue is a good physicist. We had known each other before all this happened, but we got to know each other a lot better in the course of this controversy. During this time he was chairman of the Space Science Board of the National Academy of Sciences. This is a very prestigious position and, as far as advisory boards go, it is the most powerful one outside of NASA.

Our findings upset Donahue from the very start. So did the handling of the subject by the press. *The New York Times* quoted Van Allen, the head of the Department of Physics and Astronomy at the University of Iowa, as saying that our cometary hypothesis was "pretty unavoidable." This prompted Donahue to send off an angry letter to Van Allen, thinking that he had played a role in my work. But I am not on a leash, even though some people wish I were. I greatly admire Van Allen, but he has no control over what I do and has not had any since I got my degree back in 1964.

Donahue was not the only one to contact Van Allen on the controversy I had caused. Thomas Gold, the noted Cornell astronomer, did so as well. Perhaps he, too, felt uneasy about voicing his comments to me directly and decided to do it through a close associate. The small comets upset Gold terribly. I have known Gold since I was in my early twenties and his response came as a surprise. Gold is a maverick himself. He believes, contrary to current opinion in geology, that most of the oil and gas in the Earth does not have a biological origin. To prove his hypothesis Gold is presently trying to mine oil from an area in Scandinavia that is thought to be the impact site of an asteroid. Current opinion holds that the area is lacking in the geological foundation necessary for gas and oil deposits.

But Donahue would play a much larger role in this controversy than Gold. Donahue is a very explosive personality. When he talked to reporters about me, he sometimes sounded like he was on the next horse out to get the small comets. One reporter for the *Ann Arbor News* characterized Donahue's attack on my efforts to bolster the small comets theory as "bitter." Donahue denied it. But it did not matter. I expected this. When *The New York Times* ran its first story on our findings, the reporter quoted Dessler as saying that the comet theory, if correct, was "probably the most important finding in space science of the last decade," and that it would "make or break" my reputation. Donahue wanted to hang Dessler by his thumbs. He thought it was improper for an editor of a scientific journal to use his position to promote public controversy and put the reputations of contributors to his journal at risk.

Donahue asked me to co-author a letter to *The New York Times* to straighten things out. I agreed. "We regret the public contention," our letter said in part, "that has had the effect of transforming a civilized discussion between scientists interested in arriving at a mutually agreed upon explanation for a puzzling discovery into a life or death struggle for our scientific reputations. We two, who have been portrayed as bitter adversaries in some of these stories, wish to assert that contrary to the impression that has been created, we have very high respect for each other as scientists and as human beings."

The letter was never published. That further angered Donahue. He was also upset because I had used the word "startling" in the final version of our paper. Originally, I had used the word "plausible" in describing our interpretation. But Donahue and others had kept calling it "startling." So I appropriated the word for the published version of our paper. Donahue was not happy.

Thomas Donahue is one of the leaders in the area of planetary atmospheres. Another is Donald Hunten, a professor in the Department of Planetary Sciences at the University of Arizona in Tucson. Hunten has spent a lifetime on the theory of planetary atmospheres and in particular on how water slowly evaporates from our planet. It is a very elegant, a very beautiful theory.

This is how it works. A water molecule comes up off the ocean and rises into the atmosphere, where ultraviolet rays split it into hydrogen and oxygen. The hydrogen, because it is so light, eventually flies off into interplanetary space. The oxygen, meanwhile, comes back down to Earth and oxidizes something, like iron, to form rust. That is the way you lose one water molecule. This is a beautiful theory because it explains why we have this big envelope of hydrogen gas above our atmosphere and it explains how some small fraction of the Earth's oceans has evaporated over time.

What could be wrong with that? In fact, only one thing. All the equations in the theory are based on an important assumption. The assumption is that there is no source for the water other than the oceans. But the small comets provide a source of water that amounts to one thousand times the rate of evaporation calculated by Hunten. For every gallon of water he has evaporating each year the small comets bring in a thousand gallons of water. Which means that the detailed mathematical equations which Hunten has spent a substantial part of his life working on are wrong because he does not account for this additional source of water. He did not make any mistake in his arithmetic, but in his assumption of the supply of water.

A scientist toils through life writing paper after paper that culminates in a magnificent theory. Then one day someone comes along and finds that a fundamental assumption is wrong and a large part of his life's work collapses just a few years before retirement. What do you expect his reaction to be? I understand how he feels, but there is nothing I can do about it. I feel like a spoiler. Hunten and I get along very well despite the fact that we both have very strong personalities. We openly argue quite a bit and these heated exchanges make a lot of people rather uneasy. Hunten is very intelligent, very honest, and very influential.

So Donahue and Hunten joined forces to defend their model of planetary atmospheres against this intrusion from outer space, the small comets. Call it damage control. It was not the first time they had tried to fight off the suggestion that the Earth's atmosphere has an external water source. But this time, Hunten said to Donahue, in effect, I'll let you take care of it. So even before our papers were published in *Geophysical Research Letters*, Donahue was preparing his criticism of the small comets theory.

Donahue saw "a large number of problems" with our small comet hypothesis but there was no anger in the tone of his published Comment. Compared to some that would follow it was almost kind. One of the problems that bothered Donahue was Venus. How could Venus avoid this cometary bombardment?

Where was all the water on Venus? Venus, of course, is one of
Donahue's very favorite subjects. He had become well-known for
his Venus work.

Compared to the Earth, Venus today is exceedingly dry. Very
small amounts of water have been measured in its atmosphere. It
is estimated that there is a hundred thousand times less water on
all of Venus than there is in just the Earth's oceans. Donahue
proposed that Venus once had an Earth-like ocean early in its
history but subsequently lost it to evaporation. Donahue came to
this conclusion after analyzing the data from an instrument on a
releasable probe carried by the Pioneer Venus orbiter in 1981. Just
as the probe plunged into the lower cloud layer about thirty
miles above the surface of the planet the instrument got clogged
up with what was almost surely sulfuric acid. During this time a
measure of the ratio of heavy hydrogen, or deuterium, to ordinary
hydrogen was taken. The value was quite high. It was about a
hundred times greater than the deuterium-to-hydrogen ratio on
Earth. This is the smoking gun that, for Donahue, raised the
issue of Venus's wet past.

To understand why, you have to realize that Venus and Earth
have long been regarded by scientists as nearly identical twins
separated at birth. Both planets were formed from the same
stellar material approximately 4.5 billion years ago and resemble
one another in fundamental ways, in size, mass and gravity. So
both planets are thought also to have been originally endowed
with the same ratio of deuterium-to-hydrogen. But while Venus
grew up close to the Sun and went dry, the Earth was brought up
farther out and was able to maintain a sizable ocean.

But what happened to all the water on Venus? Donahue
proposed that a once Earth-like Venus lost its oceans due to a
runaway greenhouse effect that boiled the oceans, sending much
of the water into the upper atmosphere where sunlight tore both
hydrogen and deuterium atoms from water molecules. The ordinary hydrogen would have escaped into space. But the deuterium, being twice as heavy as ordinary hydrogen, was left
behind as a residue. The loss of ordinary hydrogen would show
up clearly as an elevated amount of deuterium in the Venus
atmosphere. But there are problems with Donahue's proposal. He

makes an assumption about the original ratio of deuterium-to-hydrogen on Venus, despite the fact that this ratio varies widely throughout the solar system. He also assumes that the water abundance on Venus has been declining throughout its lifetime.

My Reply to Donahue simply said that the small comets could also account for the presence of water vapor in the Venus atmosphere. An ancient ocean was not necessary. Of course, these comets could not possibly bring in as much water to Venus as they bring to Earth. The flux of the small comets at Venus cannot be the same as it is on Earth. The Sun vaporizes most of the small comets as they approach Venus. Maybe only one out of a thousand small comets actually reaches Venus. So I said there was no need to invoke an early ocean on Venus to explain the water vapor in its atmosphere. A few small comets entering its atmosphere would do just as well.

Others had already questioned Donahue's scenario. One was Sig Bauer at the Institute of Meteorology and Geophysics at the University of Graz in Austria. Back in 1983 he authored a paper for *Annales Geophysicae* entitled "Water on Venus: Lack or Loss?" In it he had suggested a possible cometary origin for the atmospheric water content of Venus. In a letter to me a few years ago, Bauer wrote: "Your observations and interpretation are of particular interest to me in connection with Venus since I always held the (minority) opinion that Venus was formed with little or no water and the present water content came from the outside."

John Lewis had voiced the same opinion back in 1974 while at the Planetary Astronomy Laboratory at the Massachusetts Institute of Technology. In a paper published in *Earth and Planetary Science Letters*, Lewis concluded: "It is probable that Venus's present hydrogen loss rate is appreciably smaller than the rate of injection of hydrogen from meteoritic, meteoric and cometary water."

Lewis, like Bauer, had looked at the constituents of the Venus atmosphere and concluded that the water there could have been supplied by an occasional comet. Comets, whether large or small, are a much more reasonable source for the water vapor in the Venus atmosphere than the evaporation from an ancient ocean. But the proposals of Bauer and Lewis were not popular in their time. Mine is not popular today.

The core of Donahue's Comment had nothing to do with Venus, however, but with the Earth. If the small comets were bringing all this water into the Earth's atmosphere at an altitude of one hundred to two hundred miles, Donahue said, then someone would have measured it. My Reply was that these comets were not laying all this water on top of the atmosphere. You cannot stop a hundred tons of water falling in at three miles per second or more at those altitudes in the atmosphere. And once that water vapor plunges into the lower atmosphere it gets so mixed up with what is already there that the infall of new water is beyond detection at the surface of the Earth.

I think my Reply stung a little. Our atmosphere is composed of two main layers. The first layer consists of the winds that circulate near the surface of the Earth. These winds travel not just from Nebraska across Iowa but up to about seventy miles high, to a region called the homopause. The winds here are part of one big circulation cell, rather than a bunch of little ones all piled up on one another. The homopause separates the lower zone of the atmosphere from the upper zone. It too has winds in it, but they are not locked closely into the winds in the lower atmosphere.

Donahue's point was that if the small comets come in and stop before they reach the general wind pattern in the lower atmosphere, then the water should stay up there and build up to the point that anybody should be able to detect it. He cited the microwave measurements of John Olivero, a meteorologist at Pennsylvania State University, indicating that there is indeed no large amount of water up there. But if the water did remain up there, it would be obvious. There would be a worldwide ice cloud. On the other hand, if the water vapor from these objects actually plummets down through the homopause it would simply mix in with the water and clouds already present in the lower atmosphere.

Our work showed that the cloud of water from a disrupted small comet enters the Earth's atmosphere like a piston of water vapor. When this piston reaches an altitude of fifty miles, well below the homopause, it is still moving at almost twice the speed of sound. It is simply not going to stop at the top of the atmosphere. Hunten found my description of small comets as pistons of water vapor punching through the atmosphere "ingenious." But it is really quite trivial, a few calculations done on a lunchroom napkin.

Donahue also thought that the supply of water brought in by our small comets was just too large. He insisted that the infall of small comets into the atmosphere could not be continuous. If it is, said Donahue, then we had too much water coming in over geological time. Over the four billion plus year history of the Earth this would be enough water to fill the ocean basins four or five times over. That seemed to bother everybody. But who would expect the infall of comets to be exactly constant over some four billion years? Besides, it is just incredible that we were even talking in terms of ocean units in the first place. We were not arguing over a thousandth of an ocean, or a millionth of an ocean, or even a thousand oceans or ten million oceans. I said it was one ocean. Donahue said it was four or five oceans. What difference did it make? We were in the same ballpark.

Chapter 6

The Creation of the Oceans

It is a strange planet we live on. There is water everywhere. There is so much water on this planet, in fact, that it is beyond our ability to comprehend it. We do have estimates of the total world volume of water, but the numbers are beyond our direct experience. Perhaps the best way to begin to grasp the quantity of water in the world is to portray it in terms of the number of bathtubs we would need to contain it. If we take a bathtub to contain about 35 gallons of water, then the world's supply of water would fill about 10 quintillion, 566 quadrillion, 293 trillion bathtubs to the overflow vent.

You would need most of those bathtubs—more than 97 percent of them—just to hold the contents of the oceans. The rest would hold the world's fresh water supply, of which more than three-fourths is locked up in the polar ice caps. Most of the remainder lies in groundwater and soil moisture. Lakes and swamps account for a little more than a third of one percent of the world's fresh water supply and the atmosphere contains less still—just a few hundredths of one percent. But small percentages can be just as deceiving as large numbers. The atmosphere actually holds enough water to fill hundreds of trillions of bathtubs.

Current wisdom favors the view that most of the Earth's water has been here since the birth of our planet, which is estimated to have occurred some 4.5 billion years ago. It is widely believed that the Earth was formed through the accumulation of cosmic debris left over from the giant nebula, essentially a dust and gas cloud, that formed the Sun itself. The debris particles, which are thought to have contained water, eventually collected together through the force of gravity to produce our planet. The water then remained trapped in the Earth until the initial crust solidified and the gases from the interior were released in volcanic events. This outgassing, or degassing, from the Earth's mantle formed an atmosphere rich in water vapor and other gases, known as volatiles. Then, as the Earth's temperature fell, water condensed out of the atmosphere. When the rains came, the original crustal minerals were eroded and weathered, forming sediments and an initial ocean.

But some scientists have begun to doubt this view. Not until the late stages of formation was the Earth either large enough or cool enough to support water. In the early stages of formation, the Earth was nothing but a large, hot cinder that had grown from collisions with many smaller cinders. Though it was eventually large enough for gravity to prevent the easy escape of hydrogen, helium, and water from its atmosphere, the Earth still could not support liquid water on its surface until it had cooled considerably. So it seems that any water present during the Earth's formation from the cinders would very likely have escaped back into space.

The alternate view is that water was brought to the still primitive Earth in a bombardment of large comets, which are conjectured to have been quite common in those first hundred million years or so of the Earth's history. Some of this comet water would have remained on the surface and some would have mixed with the crust of the Earth and its underlying mantle and have been brought to the surface by volcanoes. Though appealing to some scientists, this scenario for water delivery through an early comet swarm is still not well-defined.

But in either case—water here at the start or brought in by large comets—it is generally agreed that the Earth probably had most of its water quite early in its history. I, too, had always assumed that the oceans were here in the beginning. I thought it had long been accepted as proven fact. I had never given it a second thought. Then one day Donahue, who had seen the early drafts of our original papers, realized that the small comets, if they were real, could have brought in enough water, over time, to supply the world's oceans.

This was the first major implication of the small comet discovery and the one that drew the most attention from the public. Initially, I was more upset than satisfied by the idea. So I did the calculation myself. A small comet falls into the Earth's atmosphere about every three seconds. Each one contains about 100 tons of water. Some 25,000 such objects fall to Earth every day, or about ten million on average over the course of a year. Some years would have more, some less. But since the Earth was formed, 4.5 billion years ago, about 40 quadrillion small comets would have fallen into the atmosphere. The total volume of water deposited by such an influx of small comets makes these objects not only a reasonable, but a likely candidate for the source of the world's oceans.

This is no magic trick. It is a simple deduction from the known facts. I will be the first to admit that it seems almost inconceivable that the ocean waters should come from *out there*. It sounds silly. One skeptical oceanographer even dubbed it "the trickle down theory of the oceans." But that is exactly what it is. I have spent some time psychologically adjusting to this rather striking disparity with my previous education. Being a very conservative person, I probably change my ideas more slowly than most.

I searched the scientific literature looking for solid evidence to back up the notion that the oceans were here in the beginning. I thought that if I could find proof that the supply of water originated early in the Earth's history, I would have to reject the notion of a constant bombardment of small icy comets. Reporters who called me on the story actually were quite helpful in this respect. They, too, searched for evidence to back up what they had learned in high school and college. And they, too, were disillusioned.

There is a lot of conjecture about what went on in the early Earth, but very little data. Our beliefs are based almost entirely on the geological record left by sedimentary rocks. And the oldest sedimentary rocks, which were formed in the reaction of crustal minerals and water degassing from the planet, have been dated at close to four billion years old. But there is no agreement on how much water was needed to produce them.

Just how quickly the oceans formed is actually a subject of some controversy among geologists. Most geologists lean toward an early formation of sedimentary rock and ocean. They think that the sedimentary material accumulated very early in the Earth's history, meaning that the water volume of the early ocean would closely resemble today's ocean. They believe there has been a kind of cannibalism of the old rock to make new rock. So that what looks like new sedimentary rock is actually old rock that has been weathered, eroded, and redeposited. Others think the sedimentary mass has accumulated gradually over time and the oceans likewise. Indeed, the proof that the Earth had as much ocean water in the past as it does today lies largely in the lack of evidence to the contrary, according to Gustaf Arrhenius, the highly respected geochemist at the Scripps Institution of Oceanography whose specialty is the early history of the ocean and atmosphere.

The small comet scenario calls for a slow acquisition of the world's waters. Adding one inch of water to the Earth's oceans every ten thousand years is undetectable. There are simply no rain gauges that accurate. In addition, no one really knows what the water levels are on Earth. If the small comets are real and we go back far enough, say a billion years, then we should find one quarter less water in the oceans. But we cannot determine that this was the case with any accuracy because the time period is so remote. Some people ask me whether the rock record shows an increase in water levels. The answer is: you cannot tell. The change in water level is just too small and the Earth's crust shifts too much.

Even Donahue has admitted that the small comets could not be ruled out to explain the origin of the oceans. "Data concerning ocean volume and sea level," he said, "are too imprecise and limited in time..." So it is all still quite problematic, this origin of the ocean waters. No one is too certain about where the Earth's waters came from in the first place or just how quickly the ocean basins were filled.

I know people do not look at scientists as innocent, but they really are innocent. And the suffering that I have endured in this affair is mainly the loss of this innocence. I assumed, like everyone else, that science was relatively secure in its knowledge of the origins of the oceans. But it turns out that most of what we know about the subject are assumptions based on our mistaken impression of a famous paper by William Rubey, published in the *Bulletin of the Geological Society of America* in September of 1951. Everyone cites Rubey's paper on the "Geologic History of Sea Water" as saying that the water must have been here in the beginning. But Rubey did not say that.

Rubey had asked himself if all the water on Earth could have come from all the leaching and weathering of rock that had taken place over the Earth's four billion plus year history. And the answer he came to was: no-way. The water has not come out of these rocks. Rubey said that by normal evolution, by weathering and such, there should be far less water, carbon dioxide, nitrogen, and chlorine, or excess volatiles as these substances are known, on Earth than actually are here. Or else, he said, our understanding of the history of the Earth is quite wrong.

Rubey thought that the water must have arrived on the surface of the Earth *over time*. He actually said that in the paper. He even briefly considered the possibility that the excess volatiles might have been gradually supplied by an extraterrestrial source. But he dismissed it, given "the exceedingly slow rate at which meteoritic debris is reaching the Earth today." He then concluded that the most likely source of volatile elements was the outgassing from volcanoes.

Rubey clearly condemned the *quick soaks*, those who prefer to have all the water here at the beginning, and came down in favor of the *slow soaks*, those who increase the water content of the oceans "by small increments over a much longer period of time." I was amazed when I read the paper. It was not subtitled "The Solution" but "An Attempt to State the Problem." Rubey had wanted merely to define the problem, not resolve it. What a loss of innocence on my part. So this was the foundation on which all our statements about the origin of the ocean waters are based. It was mind boggling. Bright man. Famous paper. But somehow misinterpreted, perhaps because it is a long paper and merely cited, not read carefully.

Besides, not everyone thinks you can squeeze the oceans out of hot rocks. The outgassing of the Earth through geologic time does not really solve the problem of excess volatiles, according to oceanographer Pat Wilde, head of the marine sciences group at the University of California, Berkeley. It is very difficult to get volatiles out of igneous or volcanic rocks. Though the distribution of some fifty-eight elements in sediments and seawater can be explained by the weathering of igneous rocks—a concept called Geochemical Balance—it does not explain the volatile content of the ocean. The numbers simply do not add up. You need another source, says Wilde, and the source may well be these small comets.

Wilde believes that the composition of the oceans may be the crucial clue to the problem of the excess volatiles. If the water was here from the beginning then it should have changed over time. When you start with a reservoir of water, it is like a fresh pool. But if you let it sit, it will become increasingly stagnant. It deteriorates because there is no refreshing of the waters. On the other hand, if you have a source of fresh water coming in continuously, then eventually the ocean should come into equilibrium with that source.

In the case of the oceans, if you start with as much water as we have now, then you would expect the salinity, for example, and the amount of dissolved materials in it to change over time. But the geologic record of the past 500 million years indicates that the salinity of the oceans has been relatively constant. This conclusion is based in part on living marine fossils that have a restricted tolerance for salinity but have been around and unchanged for about half a billion years. One example of such a living fossil is the brachiopod *Lingula*, a small invertebrate creature with a brown, tongue-shaped shell that lives in mud or sand and has a preference for quiet waters.

Wilde realizes that the small comets may bring in more than enough water to fill the ocean basins over time but he has no problem accounting for any excess. Extra water may be used to put water into the Earth's mantle and grease the motions of the continental plates, all the while keeping the actual ocean volume relatively constant. "Some of the excess water might be one of the reasons for plate tectonics," ventures Wilde. "You have to have this type of lubricant otherwise everything would lock-up occasionally." It would be "a startling geological revelation," he adds, if the composition of the small comets should mimic the composition of the oceans.

There is now evidence that this may be the case—at least for comets in general. Studies of Halley's Comet conducted by the European Space Agency's tiny Giotto spacecraft in 1986 indicate that the comet's chemical fingerprint seems to match that of the Earth's oceans. In the planetary sciences, the relative amounts of the isotopes of hydrogen, oxygen and sulfur present in a sample can serve as a fingerprint of its origins. Isotopes are the members in the family of a given element, such as hydrogen, for instance, which has an isotope called deuterium. Hydrogen differs from the heavier deuterium atom in the weight of its nucleus.

Analyses of the Giotto results carried out by Swiss physicist Peter Eberhardt and his colleagues show that the Earth's oceans contain the same ratio of deuterium-to-hydrogen as Halley's Comet. The relative concentrations of oxygen isotopes are also the same. And the same is true for sulfur. Eberhardt believes that these findings are in agreement with models proposing that the Earth may have acquired its water and other volatiles from comets. He is referring to those who believe that the Earth's waters may have been deposited in a storm of large comets early in the Earth's history, of course, and not to the current influx of small comets that I have proposed. But he certainly does not rule out this possibility either.

The bottom line is that the oceans could well have come from comets.

Chapter 7

Heat, Dust, and the Origin of Life

Alice Springs lies in the heart of Australia. Nearby is Ayers Rock, a 1,260-foot high monolith located in the center of the red Aussie desert. This natural wonder draws visitors from around the world, but in the spring of 1986 Alice Springs had yet another natural wonder on its hands. Tens of thousands of comet watchers flocked to this tiny, clear-aired town in the outback to catch a good view of Halley's Comet. But as everyone knows, the celebrated, once-in-a-lifetime visitor simply fizzled out. The comet was fuzzy and feeble and everyone was disappointed. The comet gazers had come so far, and the comet, just 39 million miles away, had come so close.

The let-down put such a somber mood on April Fool's Day in Alice Springs that a group of soldiers decided to fill a nearby trench with dry ice and announce that a chunk of Halley's Comet had fallen to Earth. They added credibility to their story by telling passersby that a group of scientists from the United States were coming to retrieve it. Word of the event spread rapidly and before long local residents were in an uproar. Americans, they said, were stealing "our comet."

I, too, was accused of stealing from the great comet. The timing of the publication of our findings irritated cometary scientists. There is certainly irony in the fact that just as our theory of the small comets attracted the attention of the scientific community, Halley's Comet so disappointed everyone else. A lot of people believed I had comet fever. They said all this was brought on because Halley's Comet was coming and everybody was obsessed

with comets. They thought I was hopping on the Halley band-
wagon only to distract everyone's attention away from it.

This was certainly not our intent. It is true that much of our
work deducing the probable nature of the small comets took
place just as scientific studies were being conducted on Halley's
during its latest little visit to the terrestrial neighborhood. But all
along our work was really totally independent of the 1986
appearance of Halley's Comet.

Water is the primary ingredient of all comets. For decades the
assumption has been that large comets like Halley's were one-
third to two-thirds water and the rest dust. But I thought the
small comets could not contain much dust. If they contained as
much dust as Halley's Comet they would be creating these
tremendous fireballs in the sky as they entered the Earth's
atmosphere. But we do not see such fireballs very often, so the
small comets could not be very dusty.

The question of the composition of the small comets was one
of the first issues we had to address. But unfortunately, thanks to
Alex Dessler, we had to address the topic repeatedly during the
course of the debate. I think that Dessler tried to bury me with
Comments related to the composition of the small comets. He
accepted the first Comment on this topic on April 29, 1986. Two
days later he accepted another Comment on composition. Less
than two weeks after that Dessler accepted a third Comment on
the topic and a month later he accepted the fourth and final
Comment relating to the composition of the small comets.

The topic was, I will admit, an important one. The small
comets had to be composed of something other than icy water.
How else could they survive the heat from the Sun as they
reached the vicinity of the Earth? These comets had to have a
protective cover, a mantle, in other words. Of course, whatever
this mantle was made of would, like the comet water, eventually
end up on the Earth and perhaps be of significance for the planet
and life upon it.

Our critics on this issue were a distinguished group. There
was David Rubincam of the NASA Goddard Space Flight Center
in Greenbelt, Maryland, and four Californians: Christopher

McKay of the NASA Ames Research Center, Donald Morris of the Lawrence Berkeley Laboratory, and John Wasson and Frank Kyte, both of the Institute of Geophysics at the University of California in Los Angeles. Their Comments were variations on the same theme.

What Rubincam and the others said, essentially, is that if the small comets were to survive their journey to the Earth, these objects would have to be protected from the heat of the Sun. This meant that the comets would have to have a mantle and most of them chose to construct this protective cover out of dust. They simply assumed that the small comets would be like the large comets and contain a lot of dust. McKay, for instance, concluded that the dust cover on the small comets would have to be about fifty feet thick to survive the Sun's heat, but this was clearly at odds with my estimate of the size of these objects—about forty feet in diameter.

Everyone insisted, Morris included, that a minuscule amount of dust would certainly not be enough to form a protective mantle around a small comet and that any such small amount of dust was inconsistent with our knowledge of the high dust content of known large comets. But even assuming you could mantle these objects with dust, they said, where is all the extraterrestrial dust that would have come from all the small comets fallen to Earth in the past? If the small comets were real, said Wasson and Kyte, there should be from ten thousand to a hundred thousand times more extraterrestrial dust on Earth than scientists have been able to find.

My critics were essentially correct. The small comets did have to protect themselves from the Sun's heat. They did need a protective mantle. But the mantle need not be dust. My critics could only think in terms of the large, known comets. They assumed that the small comets would have to have the same dust content as the large comets.

But the small comets are different and contain only a fraction of the dust of the large, well-known comets. I do not think it unreasonable to expect such a difference. After all, in a comparison of masses, a small comet is to a large, known comet what a large comet is to the Earth. Large comets are a hundred million to a billion times more massive than the small comets. Several simple scenarios could explain the different dust compositions of the large and small comets. Perhaps the large comets accrete more cosmic dust than their tiny cousins because of their greater gravity. Or perhaps they were formed in a different, dustier region of the solar system.

But current thoughts on cometary evolution do not predict small comets, McKay pointed out. He is absolutely correct. No one ever thought there might be such objects. The whole scientific literature is devoid of any real discussion of the possible existence of small comets. But I think the reality of these objects is a clear indication that the current models of cometary evolution are quite incorrect. And when I say the small comets are about forty feet in diameter, I refer to an average size. Some will no doubt be smaller, and others larger. We are not talking about two completely distinct classes of objects, the known large comets and my own small comets. I fully expect there to be cometary objects of all different sizes and different dust content cruising through the oceans of space.

Regardless of the amount of dust in the small comets, however, a dust coating simply cannot be the dress around these objects. The comets have to live longer than just a few trips around the Sun, so mere dust will not do. But water and dust are not the only cometary constituents. Organic materials are also present in comets in small amounts. And laboratory experiments by Louis Lanzerotti at AT&T Bell Laboratories, Robert Johnson at the University of Virginia, and others, have shown that if you have organics you can get a crust on a comet. In a series of experiments, they bombarded water-ice, containing such organics as methane gas, with radiations similar to those found in space. The result was a durable, thin, black crust. It was not a very thick crust perhaps, but one that should be sufficient to protect a small comet in its approach to the Sun.

With a little vaporization and a little cooking from the Sun the mantle on the comets should resemble nature's version of Saran Wrap. If you start out with simple carbon compounds and expose them to millions of years of energetic charged particles and ultraviolet radiation from the Sun and distant places in the cosmos, it is quite possible to produce a black crust about half an inch thick on a comet. This carbon coating would bind the comet together. By carbon, I do not mean what you find in a pencil. It would be a carbon polymer; the carbon, in other words, would be mixed in with other elements, perhaps into chains of molecules. So it is going to be an amorphous, perhaps flakey, tar-like substance. Such a carbon mantle would give the small comets, coming no closer than the Earth's distance from the Sun, just the lifespan they need—a hundred thousand years to one million years.

I proposed this black crust well before we learned that Halley's Comet itself had a very dark surface. It annoys people when I mention this, but it was just a natural extension of our work. Everyone had expected Halley's to look like a dirty white snowball. But the probes found that large areas of its surface were extremely black. In fact, it was described as one of the blackest things ever seen in the solar system.

The comet's nucleus was also not spherical, but irregularly shaped, like a potato, and had holes in it. Everyone thought the entire sunlit side of the comet nucleus would be illuminated. It was not. The comet's illumination came from the jets of gas and dust that are blown out from the holes in its nucleus. And these holes, it just so happens, suggest an intimate link between the large, Halley-type comets and my own small comets.

Large comets travelling into the inner solar system are known to split and flare. No one knows why. They just break apart. I thought if the small comets are really there, then no large comet could exist very long without being hit by them. So I did some calculations. I began by trying to show that the small comets were too numerous and that any large comet out there would just be blown out of existence by them in a short time. But it turns out that the number of small comets is just about what you would expect to account for the random splitting and flaring of large comets. Not only that, but it is also the right number required to put several holes in the black surface of Halley's Comet.

The photographs taken by the Giotto spacecraft clearly show the holes in its nucleus. I think Halley's Comet has been hit by a number of small comets. I did not bother to publish this finding even though no one else has been able to explain the random splitting and flaring of comets in interplanetary space. I knew that our conclusion would irritate people to no end.

I should mention that Donald Morris, to his credit, did wrestle with the possibility that the mantles on the small comets might be made of carbon, a basic ingredient in all forms of life. But, Morris argued, there are limits on the amount of carbon that comes down to Earth. If these comets are coming in, he said, then the carbon has to dribble down through the atmosphere and end up on the ocean floor. And there are measurements of carbon on the ancient ocean floor, core samples by geologists such as Wendy Wolbach of the University of Chicago, that put an upper limit on how much carbon you can bring in with these comets. And it turns out, Morris said, that the amount of carbon down here is too small, so your comets are out of luck.

The point was a good one, but incorrect. If I have small comets slamming into the atmosphere at supersonic speeds, the carbon is going to ignite. There is oxygen up there. So the carbon is not simply going to fall down as carbon. It is going to burn like a furnace. It will not even get close to the Earth. It has to pass through atomic oxygen, which is highly reactive. It will be consumed by this oxygen as it falls through the atmosphere and turned into carbon monoxide, carbon dioxide, and other carbon compounds. Very little of it will float down to the Earth's surface as soot.

So the more appropriate question is how much new carbon exists on Earth? If additional carbon can be found on the Earth, what is the upper limit per year that is not identified as already being here? Heinrich Holland, a professor of geology at Harvard, has conducted an exhaustive examination of this issue. He calculated just how much carbon exists in the atmosphere, biosphere, and crust of the Earth. He also estimated that about three-quarters of the new carbon on Earth is actually recycled carbon from the weathering of sedimentary rocks. But Holland could not quite pin down where the remaining twenty-five percent of new carbon comes from. He suggested that it might be coming from the deep interior, that volcanoes are bringing up juvenile or new carbon from the Earth's own mantle. Of course, where but down below would a geologist look for a source of new carbon?

But the amount of carbon brought in through the influx of small comets each year can also account for Holland's remaining twenty-five percent. In fact, if you take the total amount of carbon on the Earth and divide it by the current rate of additional new carbon, you come up with about four billion years. In other words, if the comets have been coming in steadily over the age of the Earth, they would not only have brought us the oceans, but most of the carbon on this planet as well. And if you go back to Rubey's paper on what the excess volatiles are, water is number one and carbon is number two. This is incredible. Of course, if it is true, then the carbon in the trees around us and the carbon in our own bodies will have come from comets as well.

Life, and its origin, is a delicate subject. Conventional wisdom holds that the key molecular building blocks of life were formed on the Earth itself. Modern laboratory experiments suggest that gases in the Earth's primitive atmosphere—such as water vapor, methane and ammonia—were broken up, either by exposure to the Sun's ultraviolet light or to lightning discharges, and then recombined to form other molecules, such as hydrogen cyanide and formaldehyde, which are thought by many scientists to be chemical building blocks of carbon-based life. But whether such photochemical and electrical reactions are indeed responsible for the overall terrestrial inventory of complex organic molecules is not known with any confidence. This uncertainty can be traced directly to the composition of the ancient atmosphere, about which very little is known.

Scientists during the past three decades have considered an entirely different scenario for the origin of life. The idea is that comets may have supplied the Earth with the organic materials needed to create life. It was back in 1961, in fact, that John Oró at the University of Houston first suggested that the biochemical molecules from which life arose on the primitive Earth may have been brought here by comets. It is a distinct possibility, as nearly a fifth of the comet's nucleus may be composed of organic matter. But many critics objected to the idea on the grounds that an object large enough not to burn up in the Earth's atmosphere would hit the ground with such force as to destroy all the organic material it contained.

Today, however, the notion is no longer wildly controversial and is the subject of ongoing study. A group of scientists at Cornell University led by Paul Thomas, along with Leigh Brookshaw at Yale University, now believe that much of the organic molecules needed to create the first forms of life on Earth could well have been brought in by comets that bombarded the planet early in its history. Their study, actually a simulation of the impacts formed by comets of various sizes, took into account their suspicion that the Earth's atmosphere was probably ten to twenty times more dense three to four billion years ago. This thick veil could well have exerted an "aerobraking" force on incoming comets of a certain size.

They found that large comets a mile or so in diameter would plunge into the Earth without sufficient aerobraking. They thought that small comets, on the order of those that I have proposed, would certainly slow down sufficiently, but would not contribute enough mass to make a difference. They concluded that the most promising candidates were comets of medium size, those measuring about 400 feet across. Such comets would have slowed down enough during entry into the Earth's atmosphere for their organic components to survive the impact. The implication of their work is that a cometary bombardment could have brought in a hundred to a thousand times as much organic material as the Earth itself would have produced photochemically during the same period.

But the small comets with their constant flux, smaller size and lower speeds do not require a thick atmosphere to deposit the initial inventory of life producing organic matter on the Earth. In fact, the initial impacts of small comets on the barren cinder that would eventually become our home could well have been responsible for producing the early atmosphere itself. Later, though the surfaces of the small comets would burn up in their descent through this new atmosphere, all organic matter buried inside them might not be destroyed. Some matter could well have floated down to the Earth on a protective cushion created by the comet's own gases. So it is quite possible for small comets to have carried to the Earth the precursors of life.

The implications were piling up. We were not just talking about bringing water, and lots of it, down to the Earth but the very ingredients for life itself. And more besides—like a good portion of the atmosphere. It seems that the small comets also address the problem of how early life, which needed oxygen to protect itself from the Sun's ultraviolet rays, could have flourished on a primitive Earth thought to have been devoid of oxygen. Some of the water molecules in these comets would have broken apart from the heat of impact and formed oxygen and hydrogen gases. This would automatically have produced a shield of oxygen to protect life from damaging ultraviolet rays.

So it may be that these small comets provided not only the chemical seeds for life on Earth, but the oxygen to protect it from the Sun, as well as the marine incubator—the ocean—in which it could grow and thrive. That, in essence, would make us all the children of comets.

Chapter 8

The Atmosphere and the Ice Ages

A strange and quite beautiful cloud can be seen over the polar regions during the summer months. It is a thin cloud, wavy or banded, colored silver or bluish white. It forms at an altitude of about fifty-five miles, in the coldest part of the upper atmosphere, a relatively unexplored boundary known as the mesopause. No other cloud occurs so high in the sky. It can be distinguished from ordinary clouds only when seen against a dark sky and illuminated by the setting sun. For this reason such clouds are called noctilucent, meaning quite literally "night luminous."

Noctilucent clouds are remarkable for many reasons. But nothing is more striking than their very existence. These clouds require considerably more water vapor than can be expected from ocean evaporation. No one knows how the water gets up there. No one really understands why these clouds exist. And finding out why has not been an easy task.

Those who explore the upper atmosphere—aeronomers, as they are known—have made little headway into the mystery. Since the 1960s rocket-borne instruments have been sent up to probe noctilucent clouds. Results indicate that the clouds are composed of ice crystals formed around meteoric dust particles. But some doubts have been cast as to whether the particles are meteoric or not. Still puzzling is their sporadic appearance. Sometimes the clouds are there and sometimes not, even though the temperature in the region is the same.

It would seem that the source of water vapor in the upper

atmosphere varies over time. And indeed, if the small comets do exist, if they are responsible for depositing the water in our oceans and for sowing the seeds of life on this planet, then it would be no surprise that on their way down they should be leaving a trace of their passage—a little more water than we would normally expect—high up in the atmosphere.

But aeronomers do not like the notion of a wet upper atmosphere. They generally assume that the amount of water vapor up there falls sharply above forty miles because air carried from lower altitudes has been dried by frigid temperatures. On the other hand, there are a number of measurements in the literature that do not show this expected dryness. These higher water content measurements cannot be explained by simple upward movement of water vapor. Aeronomers are quite correct. The water is not rising from below. It is falling down from above.

George Reid and Susan Solomon are aeronomers at the National Oceanic and Atmospheric Administration in Boulder, Colorado. They have a model which computes how much water should be present at each level of the atmosphere. They used it to show that the small comets would bring far too much water into the atmosphere, even though this whole topic of water in the upper atmosphere is a kind of frontier of knowledge. We know less about the upper atmosphere, this area right above our heads, than most people realize. This is a no man's land. It begins at about thirty miles and extends another 100 miles up. We cannot reach it with airplanes. Even balloons will not soar through this region. And it is too low for satellites; the atmosphere would quickly pull them down to a fiery death. But rockets can do the job and so can certain types of radar and laser beams.

So Reid and Solomon took their model, injected into it the water coming in from the small comets, and concluded that the resulting water concentrations in the atmosphere would "greatly exceed those that have been observed." While most measurements of water vapor in the air at about forty miles and above are on the order of just a few parts per million, when the water deposited by the small comets was taken into account, their model gave a result that far exceeded this small trace of water at these altitudes.

Reid and Solomon did not want to throw the comets out with

the comet water, however. They proposed an alternate scenario. They suggested that the water molecules brought in by the small comets would not survive the impact into the atmosphere intact. The molecules would instead be broken up into an H atom and an OH molecule, known as a radical. These OH radicals would then destroy the atomic oxygen in the atmosphere and produce the atmospheric holes in our images. Never mind that the mechanism for producing atmospheric holes cannot be related to a chemical reaction with oxygen atoms. This could only happen if the objects in our images were located at altitudes where atomic oxygen is dense, somewhere under 200 miles. But the fact is that the cometary water clouds are located above the atmosphere when they produce the holes by absorbing the light from the atmosphere.

Never mind. With such a scenario, Reid and Solomon continued, you could bring a good deal more water into the atmosphere than previously thought. The flux of the comets could then be thirty times less than what we had originally proposed. The small comets would only have to measure a mere seven feet in diameter instead of forty, and weigh three tons instead of a hundred. This scenario, they said, would be consistent with current observations of the amount of water vapor in the atmosphere. Reid and Solomon were like everyone else. They were bothered by the numbers.

No one would have complained if my proposal concerned just one small comet coming in every year. Even two or ten comets would not raise many eyebrows. But at a hundred, people start thinking, that's one every three days. Still, they might concede that. At a thousand—three a day—the conservatives turn their eyes toward you and say, I don't know about this. At ten thousand, you have half the people in the world really upset. At a hundred thousand, they look at you like a creature from another planet. At a million, you are from another galaxy. And at the ten million I have proposed, well, you are not even sane.

But there is a problem with Reid and Solomon's model. It is based on a certain assumption. They assumed that the water from the comets comes to a dead stop on top of the atmosphere. It is as if some magical mechanism stops it up there. If the water did stop there, I agree that we would have too much water in the upper atmosphere. But they are wrong. This is a small problem. There is nothing up there to stop a hundred tons of water traveling at many times the speed of sound dead in its tracks. It is going to plummet deep down into the atmosphere. So their work is invalid.

What is most surprising is that Susan Solomon pointed out back in 1982 that you could not accurately calculate the movement of water vapor in the atmosphere by considering altitude only. What you needed to do, she concluded, was to consider also the motions of the atmosphere in latitude and longitude as well. This was ignored in their calculations of cometary water in the atmosphere. But what is valid one day seems to be invalid or not needed the next. There was no point in arguing.

If it is done correctly, I think there is much to be learned from running the small comet scenario through a model of the atmosphere. It could help us understand what effect the small comets might have on our climate, on the greenhouse effect, and the hole in the ozone layer over Antarctica. We know that water has a profound effect on the chemistry of the atmosphere. We know that water, like carbon dioxide, can also produce a greenhouse effect. And we believe that water ice may play an important role in the chemistry of the ozone hole.

It is also fairly obvious that an increase in the number of small comets could cause clouds to form about fifty miles up, reflecting sunlight and cooling the Earth. Comet storms could create a global ice cloud and may have been responsible for the Ice Ages. The last one, known as the Little Ice Age, was a reign of cold that began in the 13th century and reached its severest temperatures around 1750. Scientists are still not sure what caused the Little Ice Age, but it came as a result of a decline in global temperatures that averaged about one degree. Even small changes in the Earth's climate can have profound environmental effects.

Could small comet storms be responsible for the Ice Ages? It is impossible to tell with our present knowledge. We will not know just what effect the small comets have on our atmosphere until the influx of water they bring in is inserted into one of the computer models available to us. I have been trying to do that for several years. There is a place called the National Center for Atmospheric Research (NCAR) in Boulder, Colorado, where

Raymond Roble has designed and runs on a large computer one of the world's best models of the upper atmosphere. I have been there several times trying to convince him to see what happens when you inject all that comet water into the atmosphere. I would do the problem myself if I could, but without access to this model it would take a lifetime's worth of work for several people.

Roble could think I am daffy. He tells me he will do it, but four years have passed and I still have been given no results. NCAR is supposed to be of service to scientists. The government puts a lot of money into it. But perhaps Roble is feeling the pressure. Such pressures are not uncommon in the scientific community. The word gets around quickly and certain factions in this community can make life very difficult for you. All that attention can be disconcerting and dangerous, especially if you end up with positive results and are foolish enough to report them.

The small comets proposal was not the first to challenge aeronomers. The problem of a wet upper atmosphere has come up again and again. Back in the mid-1970s a group of East German scientists from the Institute of Electronics at the Academy of Sciences took a great number of density profiles of the upper atmosphere using instruments aboard the Intercosmos satellites. Their results were puzzling. Sometimes, at altitudes above sixty miles, they found a very effective absorber of ultraviolet light. They concluded that this could only be due to the presence of nitrogen oxide or water vapor. Most arguments spoke in favor of water vapor, although such a conclusion led to much higher concentrations of water vapor than are permitted by models of the upper atmosphere. Some concentrations were surprisingly as high as a thousand parts per million. They were never happy assuming that all this water had to come from down below.

More recently, Klaus Grossmann, a physicist at the University of Wuppertal in West Germany, conducted an impressive series of rocket-borne measurements of water vapor in the upper atmosphere using an infrared light detector. Most efforts to observe water concentrations at these altitudes by rocket-borne instruments are stymied by problems of various kinds. The most frequent problem is contamination. Rockets carrying experiments designed to measure high altitude water content often end up inadvertently carrying the water up with them, making their results useless.

But apparently Grossmann had designed his experiment in such a way that contamination was not possible. It was a very carefully done rocket experiment. Three times he had flown his instrument aboard sounding rockets launched from northern Sweden and three times his samples indicated there was more water up there than could be accounted for by evaporation from the oceans. The launches all took place in the winter and in the winter you expect the water vapor to remain fairly constant with altitude. The water is supplied from below and there is no loss because there is no sunlight to fragment the water molecules.

Each time Grossmann found an increase of water vapor with an increase in altitude. There was, in fact, two to three times more water in the upper atmosphere than expected. So either the water was coming up from below in a way that current evaporation theory cannot account for, or the water was coming in from outside, from an extraterrestrial source. There was no way for Grossmann to tell if small comets might be responsible for his results, but I certainly think they are.

I believe there is more water up there than most aeronomers would care to admit. But my critics were mistaken in thinking that the water from the small comets would just lie up at the top of the atmosphere. Certainly, a small amount of the water gets left up there but most of it barrels down through shear momentum and mixes in with the water present much lower in the atmosphere. If I am correct about this, then the last chance to trace the hot water vapor cloud as an object would be in the middle atmosphere, at altitudes about forty miles above us.

Luckily, it is possible to monitor the presence of water vapor in the middle atmosphere. Such measurements began in the 1950s using aircraft, rockets and balloons. But the results were inconsistent in part because water samples were contaminated by water brought up by the instruments themselves. Later, remote sensing techniques were developed that had the distinct advantage of removing the uncertainty of what was instrument water and what was atmospheric water.

Since the late 1970s ground-based microwave radiometers have been used to determine the amounts of water in the middle atmosphere. A radiometer is essentially a telescope that is sensitive to microwaves instead of visible light. These microwaves are basically high frequency radio waves much like the ones used by your home microwave oven. Because water vapor absorbs and emits at these frequencies, you can cook food in a microwave oven. When your oven pumps in microwaves, the water that occurs naturally in food absorbs that radiation and heats it up.

What microwave radiometers have found is that water concentrations in the atmosphere reach a peak at an altitude of about forty miles and decrease very rapidly above this height. But, of course, if cometary water vapor clouds do exist, these instruments should be capable of detecting large, momentary bursts of hot water as the cometary material descends and encounters an increasingly dense atmosphere.

Virtually the only people in the world who have spent long periods of time looking continuously at water vapor concentrations in the middle atmosphere are John Olivero and his group at Pennsylvania State University. They are the best in the business. Olivero is a professor in the Department of Meteorology and the instrument he uses is the Penn State radiometer, a horn antenna which can collect microwave radiation. It is a fixed instrument, meaning that it always looks at the same spot in the atmosphere. A single reading for the presence of atmospheric water vapor takes twenty minutes to collect. But the signal from a single reading is so weak that thirty-six of them are grouped together to give a single measurement for water vapor in each twelve-hour period.

Early in the game Tom Donahue cited the work of Olivero. Microwave radiometers used to monitor water variations where your objects are supposed to come to rest, Donahue said, do not see anything. They should see an increase in water vapor concentrations at least during the time it takes the objects to pass within view of the radiometer. Donahue checked with Olivero to see if these increases occurred. They did not. But Olivero recognized that if these things were plunging into the middle atmosphere, the event would not last long, just a matter of minutes. Olivero, however, was averaging his measurements over long periods of time, hours and days. If a quick event did occur, it would be lost in these averages.

So Olivero decided to go back and look at his data again. He would break the data apart, look at each twenty-minute reading separately, and compare them to the long-term averages of which they were a part. Any momentary increase should show up in the individual measurements. He did not expect to find any-thing. He did it, he said, for one reason. He wanted to "bury" the small comets. He was going to resolve this issue once and for all. "We thought Frank's hypothesis was very interesting," Olivero said at the time. "And regardless of whether it was true or not, it made us all reexamine our assumptions about physical pro-cesses going on at these levels of the atmosphere. That's a very healthy thing to do."

It would take a long time to look through so many observa-tions. I did not know it at the time but Olivero had assigned the search to a graduate student named Dennis Adams. It would be his master's thesis. Adams, being in the Air Force, was the right choice. The Air Force, and a lot of defense agencies, have an interest in scientific projects of this kind. Rapid fluctuations in atmospheric water content at high altitudes could potentially affect the detection of airborne intruders, as some defense systems are designed to spot aircraft and rockets by looking for water vapor from their exhausts.

Olivero intended to dispose of the small comets. That was good. I appreciated that. That is exactly what is supposed to be done in science. Olivero is a good scientist and a solid experi-menter. He had a lot of courage. I have a lot of respect for him. Any result he came up with, I would believe.

I heard nothing more about the project until I was thumbing through the summaries for talks being presented at the May 1987 American Geophysical Union meeting in Baltimore. I saw that Olivero had a paper with someone named Dennis Adams. A third co-author was listed as Charles Croskey from the Depart-ment of Electrical Engineering at Pennsylvania State University. The paper was on their search for rapid variations of water vapor in the upper atmosphere. No results were given. The abstracts for the schedule book need to be submitted so far in advance that the results of a particular study are often not known when the abstracts are written up.

When Sigwarth and I arrived, there were a lot of people in the meeting room. I knew who Olivero was, but he did not know me. He was getting some water to drink and accidentally spilled it all over me. There seemed to be a lot of nervous people around. I began to think there would be more to this than someone getting up and saying they did not see anything. I took a seat in the front row. When the next paper was announced I saw this young man walking up to the front who I thought might still be in high school. It was Adams. People were surprised.

Adams shuffled his viewgraphs around for awhile and began with a description of how their instrument worked, how they received the microwaves, and how they analyzed their data. There was no hint of the conclusion to come. He then explained that they had examined 390 hours worth of microwave measurements one by one for fast variations. It was a lot of work. Adams then paused and said, "Now I'd like to talk about the results." But before he put up his next viewgraph he said, "We found something."

I thought half the people in the front row were going to have a heart attack. I have never seen anything like it. Adams then showed us the events themselves. There were not many, but each was clearly very real and not noise. Each of these twenty-minute events showed a large increase from its adjacent twenty-minute periods. That seemed to indicate that sometime during this twenty minutes the water vapor suddenly appeared then disappeared from that part of the atmosphere. The short duration of the events was their most telling feature.

Adams then walked off and sat down. I noticed Olivero was sweating a bit. It took a lot of courage on their part. If you do not have much funding to begin with, and Olivero did not, you run the risk of people ridiculing you for doing something like that. A reporter once came up to me and said, "You seem to be outnumbered ten thousand to one." I replied, "I did not know the odds were that good."

Adams found four cases of "extreme variability" in the 390 hours of data collected between February 22 and March 26, 1986. At a later American Geophysical Union meeting he presented another paper in which these four events had been thoroughly scrutinized for any signs of error. They were not sporadic events due to instrument failure or errors in data analysis. These were real blobs of water coming into the middle atmosphere. They could not be impulsive injections from the surface, from the oceans, moving up. They were not due to volcanic events spewing water vapor into the atmosphere. They had to come from the top down and the only way to do that is with an extraterrestrial object.

All that remained to be done was to calculate how often these events occurred and how much water each one contained. These numbers would then have to be compared with the known frequency and the estimated water content of the small comets. Adams wrote his master's thesis on this subject. He calculated that the average time between these impulsive events was a little over four days. The expected frequency of comets for the area of the atmosphere viewed by their radiometer was about one every day and a half. The average rate we had calculated for small comets based on the appearance of atmospheric holes in the Dynamics Explorer pictures was about twice as much. This difference fell well within the variability of cometary events themselves, which we had observed to change by a factor of ten over a three-month period.

Adams also estimated the number of water molecules seen in these impulsive bursts. He figured that the smallest size for these water bursts had to be about ten tons. So the amount of water present in these events was in the same league as the small comets themselves. Adams was convinced there was a causal relationship. "Observed, short-term perturbations in microwave spectra," he concluded, "are most likely the result of a temporary increase in the water vapor concentration of the upper atmosphere from extraterrestrial 'events'..."

John Olivero was less certain. But I have no doubt, knowing the kind of scientist Olivero is, that Adams's thesis can be counted on. The examining committee of faculty members must have thought so too or Adams would not have gotten his degree. Adams has since returned to the Air Force. His thesis, unfortunately, is buried in the library of Pennsylvania State University and no part of it has been published in a scientific journal.

But Olivero would not let me down. Two years later I would hear from him again. I have some results, he wrote in March of 1990. Another graduate student on leave from the Air Force had analyzed the four remaining years of data and had found another 111 "water events." After eliminating "all other reasonable sources for the observed signals," they had concluded that these water events were most likely the traces of the passage of small comets through the atmosphere.

Chapter 9

A Masquerade for Radar

Fred Whipple is widely regarded as the patriarch of cometary science today. He was born in Iowa in 1906 and was the first to suggest, back in the 1950s, that the large comets might resemble dirty snowballs. But when the subject of my small comets came up in 1986, it infuriated him.

"If his comets have all this dust in them that real comets have," Whipple said at the time, "then our radars should be picking them up like mad. Imagine, things of that size! My God, our radars can pick up things the size of basketballs most of the way to the Moon. I don't see how they could be missed. He must have them made of pure ice."

Whipple is, of course, quite correct. The small comets cannot have a lot of dust in them, as we have already noted, otherwise we would see these tremendous fireballs across the sky. The small comets are largely water snow with, most probably, a thin carbon mantle. Can our defense radars pick up such objects before they hit the upper atmosphere? It was a good question. A lot of people had been asking the same thing. Even if these small dark objects escaped visual detection, it was thought that radar, at least, should be able to spot them.

But the question is not easily answered. To begin with, the cores and mantles of the small comets are not made of the highly reflective material that radar depends on to track airplanes. The radar signal from a small comet must be weak and must mimic that of a much smaller metallic object. This is entirely within the realm of possibility. It is this type of confusion that has enabled the Pentagon to design a large bomber, known as the B-2, which some say resembles a mere insect on a radar scope.

The early warning system used to defend the United States from airborne attack consists, in part, of large arrays of over-the-horizon radars. It is one of the best radar systems in the world. But the data this system produces is all classified. From time to time we hear of reports of strange objects in their field of view. But even scientists are not able to penetrate the security around such events. I have tried. Supplying us with that kind of information would reveal the characteristics of our defense system and would compromise national security. I believe, nonetheless, that our new radars are sensitive enough that an occasional small comet might be buried within all their data. I would also expect the defense establishment to have little interest in such objects, however, as their paths are from deep space and not from a launch pad in another part of the world.

Reporters trying to pry some information from the North American Aerospace Defense Command (NORAD) found the agency closemouthed about its ability to pick up a forty-foot chunk of snow and ice. "Our radar is infrared," Kay Cormier explained in 1986. Cormier was working for NORAD Media Relations at Peterson Air Force Base in Colorado. "It intercepts the heat from missile launches. That's why our satellites are infrared and pick up heat. We can pick up forest fires, even garbage-can fires, but not ice. Ours are not geared for that. Unless it's a man-made object we wouldn't pick it up. Our satellite sensors are geared solely for man-made objects. I've never heard of a radar that picks up ice. Our tracking system is solely for missile detection and for satellite tracking."

Wes Bodin, of NASA's Goddard Space Flight Center in Greenbelt, Maryland, is familiar with the capabilities of the NORAD system and was able to provide a few more details. "NORAD has a grouping of radars sort of like a fence," he explained. "They are broad in one direction but not in the other, so that things that cross it will show up. If an object doesn't hit in that area, then nothing will show up. Now, NORAD radar is very sensitive: it could get a signal off one of these comet-like objects. And they probably get stuff that probably looks like noise."

The ability to track a small comet-like object depends on the sensitivity of the radar. Low-gain, broadband radars, like those used by airports, track objects that are close by. Such radars do not have to be precisely aimed at a target to pick it up. On the other hand, the higher the gain of the radar system, the smaller the target it can track, but the greater the need for the radar to be pointed precisely at the target. So a high powered radar has a fairly narrow beam that requires precise pointing information—or a lot of luck—to find what it is looking for.

Snow and ice can reflect radar waves but the radar has to be able to point to it. Comets move fast. To track them you have to be looking for them. Large comets can be tracked on radar. But their trajectory is known. "If you know an object is coming and you know its trajectory, you ought to be able to track it with a high gain radar," Bodin said. In other words, comets might occasionally cross a radar's path, but no one would be likely to know what they were unless they were specifically looking for comets in the first place.

Perhaps small comets are being picked up by radar disguised as some other objects—such as meteors. It is well known that at least several billion meteors streak through the Earth's atmosphere every day of the year. No one argues this point. Most meteors are mere specks of dust, are largely devoid of water, and weigh but a tiny fraction of an ounce. Only a few of the larger ones are bright enough to be seen at night with the naked eye.

With the introduction of radar in the 1940s scientists were able to greatly expand our knowledge of these celestial visitors. When a meteor rips through the upper atmosphere, its molecules, and the molecules of the air it passes through, are torn apart to form a short-lived trail of electrically charged ions that will reflect radar signals. Radar has a big advantage over visual observation when it comes to the study of meteors. Unlike a human being, a radar can track meteors during daylight hours. And at night, an observer might expect to see only a few of the objects that a radar viewing the same sky can sense.

I now believe that a tiny fraction of these radar meteors, as they are known, may actually be small comets. Because most meteor surveys conducted by radar simply tally the observations of ionized gases left by the impact of extraterrestrial objects with the atmosphere, I would think that a small rock is probably indistinguishable from a hundred-ton snowball. We cannot interpret radar signals very well. More often than not, when we get a reflection on radar, all we know for certain is that something has passed through the atmosphere.

I thought that if a link did exist between the influx of small comets into the atmosphere and the appearance of meteors on radar, then the rate of atmospheric holes in our satellite images should show the same variations in time as radar meteors do. It turns out that meteors—like showers in April, flowers in May, and weddings in June—do show seasonal as well as other variations in time. But did the small comets show such variations? Finding out was not an easy task. It is extremely difficult to observe the same area of the Earth's sunlit hemisphere for an extended period of time with Dynamics Explorer. But we found that one such period of observation did occur soon after the satellite was operational. This period took place between the 1st of November 1981 and the 20th of January 1982.

At first we thought that the rate at which these holes appeared in our images would increase along with the number of bright meteors in the sky. So we counted up the number of holes per day over those three months and compared it to the dates of the meteor showers that occurred at these times—the Northern Taurids, the Leonids, the Geminids, the Ursids, and the Quadrantids. Much to our surprise, we found that the occurrence of the holes had nothing to do with meteor showers.

But there are two components to the meteor rate. One is the background rate, which varies slowly over days and weeks and is continually present. On top of this rate are the showers which usually last just a few days to a week. These showers occur when the Earth passes through a stream of debris created by the breakup of a large comet or an asteroid. Because the meteor rate is thought to be similar from year to year, we searched for a study of radar meteors that had occurred over the same months as the data we had collected on atmospheric holes with Dynamics Explorer.

We found a study of radar meteor rates for June 1955 through August of 1956 conducted by two Canadian radiophysicists. When we eliminated meteor showers from their data and looked only at their background meteor rates for the months of November through January, we discovered that the rate declines after the first week of November, falls to a minimum in the middle of January, and then rises again afterwards. We then plotted the rate of the holes occurring in our satellite images and found the same variation. There was a maximum at the beginning of November, a decline to the end of the month, followed by a fall to a minimum in mid-January, and an increase afterwards. The similarity between the background meteor rate and the rate for atmospheric holes was stunning.

This would suggest that the small comets are distributed spatially about the Sun in the same manner as a certain population of meteors tracked by radar. The problem is that there are far more radar meteors than small comets. During the November 1955 through January 1956 time period there were about a hundred times more meteor events as there were atmospheric holes or small comets during November 1981 through January 1982. So while small comets are likely to be related to these radar meteors, they are not one and the same. But many radar meteors could well be the debris from the break-up of some of the small comets in interplanetary space. Little pieces from the mantles of the small comets would not produce sufficiently large clouds of water vapor to be picked up as atmospheric holes by the satellite but they would certainly be large enough to be picked up by radar.

I felt that these temporal variations in the occurrence of the small comets should be widely known. Anyone attempting a search for these objects should know that the comets were ten times more likely to be seen at the beginning of November than in the middle of January. So in June of 1986 I prepared a paper on the topic and submitted it to *Geophysical Research Letters*.

The reports that came back from the referees were very critical. I think it was mostly an emotional outlet for them. This can happen when things are done anonymously. Two of the referees never understood what was going on in my paper on time variations. They were talking about one thing. I was talking about another. Only one referee stuck to the point. He insisted that observations of geophysical phenomena—whether atmospheric holes or meteors—were all modulated by the Earth's rotation on its axis and its revolution around the Sun. So any similarity of daily or seasonal behavior was unlikely to imply any deep relationship between such phenomena. Even so, he was of the opinion that neither the daily nor the seasonal variations in the occurrence of the atmospheric holes were consistent with meteor observations.

I could have answered the referee, revised the manuscript, and pushed for its publication. But Dessler was inflexible. In August of 1986 he informed me that the small comets hypothesis was not gathering support. He also felt that the journal was already providing an adequate forum to defend our findings. This debate would have to be fully aired before he would allow us to publish anything else on the subject that was not in the form of a Reply to someone's Comment. He had already been burned because he had published my first two papers. He did not feel that it was appropriate to go against the advice of the referees once again. This was as far as he would go. So Dessler rejected my paper.

Ultimately it did not matter to me. Some time later I was able to insert these findings in a Reply that Dessler would publish anyway. More importantly, I now knew the answer. I understood how the occurrence of small comets in the Earth's atmosphere varied over time.

The billions of trails left by meteors as they flash through our atmosphere are now being enlisted by the Pentagon for relaying messages in case of a nuclear war. The Defense Communications Agency and the Air Force recently completed an emergency communications network that, unlike conventional communications links, would not be disabled by the electromagnetic pulse created by a nuclear explosion. This network exploits the impact of individual meteors on the atmosphere, or the continuous background of many overlapping impacts, to reflect radio signals from one part of the Earth to another. The technique, called meteor burst communications, uses high-speed equipment controlled by computers to first sense a meteor trail and then transmit a "burst" of highly compressed information during its split-second existence.

In 1985 Billy Ficklin had been conducting experiments for the United States government on meteor burst communications using a radar installation in Alaska. Ficklin is a senior research engineer in the Special Communications Systems Laboratory at SRI International in Menlo Park, California. Certain anomalies had cropped up in his data, and after hearing about the small comets in 1986, Ficklin came to believe that these anomalies might have been caused by the presence of small comets or some other unidentified objects.

Between September 21st and October 25th, 1985, Ficklin had observed sixty large meteor scatter signals that lasted more than thirty seconds. These were larger and of longer duration than the meteor bursts that commonly occur. He suspected that these events could be created by the type of comet we had described. He may be right. The disturbances Ficklin reports are at about the right rate and at the right sort of scale you would expect from the small comets. His findings suggested a favorable prospect for conducting a radar search for these icy intruders in our atmosphere.

It took more than two years after the discovery of the small comets for someone to initiate a search dedicated to finding these objects with a radar. It took place during the first week of November, 1988. It was not done by an American, but by a Frenchman named Gérard Caudal working at the Centre de Recherche en Physique de l'Environnement. I had met this young man at an international scientific meeting in Helsinki, Finland, the previous summer and we had discussed his future search for small comets. He was quite eager and I was under the impression that he was looking for a quick way to make a name for himself.

Early in 1989 he submitted a paper on his radar search to *Geophysical Research Letters*. The paper was entitled "EISCAT Finds No Small Comet." EISCAT, which stands for European Incoherent Scatter radar, is a powerful instrument whose transmitter is located near Tromsö, Norway. Caudal had sent up radar pulses, looking for echos reflected back from the small comets at a distance above their break-up altitude. He assumed that anything moving faster than three miles per second would be a comet rather than a satellite. Caudal expected to see six comets during his search, but had found none.

I was asked to review his paper. At first glance, Caudal appeared to have done a fantastic job of eliminating the small comets with his radar search. But on closer inspection, I found that he had made a couple of crucial mistakes. The most important one was that he had looked in the wrong direction. It was difficult to believe that he had made such an error because his paper cited the three papers in the literature that told him where to look. But he made the assumption that the distribution of the small comets was isotropic, in other words, coming into the atmosphere from all directions. He did this despite the fact that the literature specifically stated that the objects were streaming past the Earth in a preferred direction. This mistake had allowed him to make the claim that he had not found the small comets. Caudal had failed to correctly cite the literature and that is something you just do not do.

My review of his paper was not a pretty sight. Not only had Caudal looked in the wrong direction, but he had also changed the character of the small comets, making them more reflective than they actually are, so that his radar would have a greater chance of seeing them. He did not know how much signal the small comets would reflect. He assumed that their surfaces were bare water snow, though the literature plainly states that these comets are mantled to suppress vaporization. When he revised his paper, he again incorrectly referred to our published work, assuming that the composition of the small comets was the same as for the large comets. I began to wonder if he had even bothered to read the journal articles on the topic.

Caudal never admitted that he had looked in the wrong direction. But he should have said so in his revision, despite the embarrassment. He claimed instead that he had looked in the right direction about half the time. I eventually had to draw the simplest of diagrams showing he never did at all. Caudal was overeager.

When you are young, science can be a very risky business. You have no established reputation. And failing to analyze your results properly can be a deadly sin. I feel sorry for young people who jumped into this fray thinking it was a child's game.

The importance of looking in the right direction cannot be

underestimated. It would also be of considerable help to anyone looking for these objects with a telescope. But first I would have to clear up all the misconceptions people had about just what the small comets would look like in the night sky to the naked eye.

Chapter 10

How to Spot a Small Comet

No one thought for a minute that the impact of a hundred-ton comet would be invisible. Meteor observers can see the arrival of an object no larger than a grain of sand. So everyone thought that you should be able to walk out in your yard at night, look up, and see the small comets blazing across the sky like burning chariots. Alex Dessler believed that too. He thought we should be able to see many of these objects every night. A few astronomers wrote to me saying that if there are ten million of these things a year the sky should be afire with them. These thoughts were not based on any calculations, but on intuition. How could we miss these objects? How could we not see them? But if anyone is to blame for these mistaken impressions, it is I. I said there were ten million house-sized objects falling into the atmosphere each year and somehow that just incited people to riot.

No one had the right perspective. No one ever bothered to compute how often you would see these objects, how long you would have to sit outside in the dark until one was within your field of view as it crashed into the atmosphere. The proper calculations show that you can indeed see them with your naked eye but you might have to stand out there a good long while to see just one. If I were to go in my backyard, it would take at least a hundred hours of clear night viewing before I would see one of them. The problem is that most of us do not go out and stare at a clear dark sky for a hundred hours in a whole lifetime. And even then, what you would see is a peculiar little streak, no sky afire.

We do not see the small comets all the time because the twenty or so that hit the Earth every minute are scattered around the planet and an individual's view of the sky is quite small. A camera on a distant satellite orbiting the Earth has the advantage of seeing an entire face of the planet at once. With such a large field of view it is able to record almost half of all the comets falling into the atmosphere at any given moment. And our satellite sees them all the time. Their entry into the Earth's atmosphere has been more or less the same since 1982. We have tens of thousands of them. Practically every image shows the vapor clouds from the comets, even in January when they are much less frequent.

For a ground observer, not only is the field of view greatly limited, but so is the viewing time. To begin with, these small comets are not visible from the ground during the day. Sunlight vaporizes these objects at high altitudes and transforms them into clouds of water vapor. Though incoming comets may generate an increase in the visible light the atmosphere normally emits, the difference is too small to notice. You simply will not see these objects glowing up there during the day.

At night, the situation is quite different. The comet may not necessarily dissipate into a cloud of water vapor before it enters the Earth's shadow. Without solar heat to vaporize it, the comet remains a compact object much longer than it would during the day. Most of the slowing down at night occurs in its last dozen miles, where the atmosphere thickens considerably. So at night you are not going to have this large cloud of dimly glowing gas but a much smaller object plowing through the atmosphere. It will come in and hit the atmosphere the way a meteor does. But because it is made of water snow it will neither glow as brightly, nor penetrate the atmosphere as deeply, as a rock.

All studies done on cosmic matter falling into the Earth's atmosphere involve dust particles and iron meteorites only. (Meteorites are meteors that you find on the ground.) An incoming particle hits the atmosphere and produces a flash. You see these things all the time. Most of them are iron and dust particles. These objects are little. They may weigh just a fraction of an ounce, but they glow. The metallic ions in these objects are very efficient at producing light seen by the human eye. What happens is that the electrons in these metallic ions are easily excited by collisions with atmospheric molecules. Then, when they relax, which they do rapidly, they emit light. Anything that contains a little bit of metallic ions glows brightly. But there are light bulbs and there are light bulbs.

Just how much a rocky or metallic object glows when it falls from the sky was worked out in the 1920s by an Estonian astronomer named Ernst Öpik. Öpik devised the double count method of tallying meteors, in which two observers view the sky at the same time. Öpik spent a good part of his life looking up at the night sky, estimating how fast an object was moving, how bright it was, and how long you could see it. These observations led to his Tables of Meteor Luminosities, which tell you its brightness, given the weight of the object, the speed when it impacts, and the source of the light—whether it is stone or iron. Öpik's understanding of the effects of atmospheric resistance and heating on meteors would later prove to be of great importance in connection with the design of nose cones and heat shields for ballistic missiles and spaceships.

Öpik's tables make it clear that just a fraction of an ounce of iron is pretty visible in the night sky. So if you scale it up to the size of the small comets, to three million ounces, let's say, then these objects should be tremendously brighter than that little chunk of iron. Those are the arguments the critics were using against us: We should be seeing extremely brilliant fireballs every thirty minutes or so. But a hundred-ton rock hitting the ground would be an unforgettable event, even if you were miles away from the impact. If that is the interaction with the atmosphere we can expect from a hundred-ton comet, then we are history. We probably would not be around if there were that many incoming, hundred-ton, rock-like objects falling down on us from the sky. At best, we would be living in caves.

But these comets are not rocks and do not behave like rocks. They contain little dust and lack the iron and other metals necessary to glow like a neon sign. And because they fall apart and vaporize so much faster than a rock plunging through the atmosphere, they can be much more massive. So I could not use Öpik's tables to calculate the brightness of incoming small comets. I looked for a Table of Snowball Luminosities but found that none exists. No one had ever worked out what happens when a soggy waterball falls through the atmosphere at very supersonic speeds, how it would glow, how bright it would be, or for how long. I would have to work it out myself.

I began looking for these answers late in 1986. A comet, I thought, would have to glow. It slams into the atmosphere at supersonic speeds, pushing and heating the air in front of it. At an altitude of about seventy-five miles, the pressure created by ramming the atmosphere equals the strength of the comet's mantle and the object is disrupted. But there is no explosive burst of visual radiation just yet. The water snow then vaporizes rapidly. This is followed by an expansion and deceleration of this blob of hot water vapor. Momentum carries it down to an altitude of about fifty miles, at which point its speed is reduced to less than one mile per second and the gas begins to cool. It is at altitudes of sixty-five to fifty miles where the gas glows brightest, though only a fraction of its energy is converted into visible light. What we see is thermal radiation, like the glow that comes off a hot stove.

But just how much glow do the small comets produce? We wanted to know whether they would be too bright. Because if they were as bright as the Sun or the Moon, we could just write them off. We needed to know the answer. The purpose of this exercise was to find something inconsistent. It is much more decisive to find one thing wrong, than to find a hundred things right. All along it has been my stated intention that if I could find something wrong with the small comet proposal, I would be the first to say what was wrong with the proposal and withdraw it.

The nose of a supersonic jet slamming through the atmosphere glows by compressing and heating the gas in front of it. This is good to know if you want to create a defense system. But it is also good to know if you want to calculate the glow from a soggy comet. In these days of ICBM missile heat shields, there is considerable experimental data on how much glow an object produces as it plows through the atmosphere. There are tables on this phenomenon. In Öpik's time it was not possible to even discuss how bright a small comet would be because nobody knew how bright atmospheric gases would glow when they are very hot and compressed. The numbers are now available to us, but calculating the glow is an immensely difficult task. Even to get a rough estimate required the use of a computer. So I asked Kent Ackerson to tackle this problem. He is a member of our research group at Iowa and was then working on an instrument that would be flown to Jupiter on the spacecraft Galileo. Ackerson is a very competent scientist and brilliant with computers.

By early 1987 we had our answer. Ackerson had come up with a surprising result. The glow produced by a small comet would be equivalent to the glow in front of a plane going about thirty times the speed of sound. That glow is roughly equivalent to the brightness of the planet Venus.

But the small comets are strange objects. They appear as bright as Venus for only about two seconds. That is how long it takes them to fall from an altitude of about sixty-five miles to fifty miles, when they slow down and snuff out. People think the small comets will hit the ground because they start out bright. At first, they do look like a big rock. But they do not go anywhere near the ground. They break up and vaporize far up in the atmosphere. It is strange for an object from interplanetary space to stop glowing that high up in the sky. A rock should glow all the way to the ground. But this is not a rock. It is a snowball.

So every hundred hours or so, you should see something in the night sky that is about the brightness of Venus for about two seconds before it vanishes. The question is, are there such objects? The answer is yes. They are objects that were never explained before and are known as Öpik's dustballs. The dustballs were never explained because they do not behave like rocks. A rock should burn its way down through the atmosphere. Dustballs start out like a very bright rock, but travel only a short distance, then snuff out.

Öpik's dustballs were known about long before the small comets. He thought the dustballs were just a big conglomeration of dust particles, big enough to cause a large glow but so weak as to disrupt and not continue down into the atmosphere. Some people called the dustballs "dirt clods." But no one could ever capture one because they blinked out long before they got to the ground and nothing was ever found at the place they were expected to hit.

Öpik never could figure out how all the dust particles in a dustball would manage to stay together. But I think the answer is water-snow. A small comet of water-snow impregnated with minor amounts of cosmic dust should emulate the visual signature of these dustballs. I think the small comets look just like Öpik's dustballs. They should have the signature of this very peculiar kind of meteor.

Öpik believed that many visual meteors were in fact dustballs. To confirm this notion, we had to establish that some ten million objects as bright as Venus were indeed being seen in the Earth's atmosphere each year. We went to a book of lists for how often visual meteors appear and looked up the number of objects with the brightness of Venus that appear annually in our skies. The number is close to the predicted number of small comets. It is not one, one hundred, or a hundred million. It is on the order of a million to ten million.

So it seems we must discard the conventional wisdom which holds that all meteors are due to stony or iron objects. Some of them are small comets. One person was quick enough to catch this. His name is Fraser Fanale. He is a planetary scientist at the University of Hawaii in Manoa. He told me once that the small comets should have the signature of Öpik's dustballs. He was right.

To observe a small comet, you must stand out on a clear dark night until you see a short bright streak that quickly snuffs out. It looks like it might be a big rock at the beginning but then winks out like a candle. You will have to be out there for a hundred hours or so to see one. A hundred hours of clear night viewing does not happen often in the average lifetime. It depends on whether you like the outdoors or not. Most of us spend our time indoors, drinking a beer or two, not watching the night sky. But even if you do, you are only going to see a few of these in your lifetime.

I think I may have seen one or two myself.

Chapter 11

Flying Saucers and Other Strange Events

The small comets hit a responsive chord. We received a flood of mail on the subject from doctors and businessmen, writers and children, teachers and preachers. Some wanted additional information, others wanted to share their own wild ideas. One geologist, for example, thought the spots in our satellite images might be produced by the impacts of large meteorites in the ocean, splashing water into orbit. But most people seemed to think we were on the right track and wanted to know more about the possible effects and consequences of such an infall of cosmic snowballs. An eleven-year-old boy wondered if germs in the comets might have killed off the dinosaurs. And a letter-writer who claimed to be "illegally confined" asked if my proposal had been responsible for the University's loss of a twenty-one million dollar contract with NASA. It was not.

A number of people wrote to tell us that they had actually seen the small comets. Many of these letters came from amateur astronomers, a fact which lent considerable credence to their stories. No one spends more time looking at the sky than amateur astronomers. They probably know the night sky better than anyone else in the world. They are the million eyes looking at the sky for uncountable hours, and they do see strange things.

John West of Texas, is one of these amateur astronomers. He has kept a keen eye out for meteors during the past two decades and wrote about a couple he had seen that were clearly "out-of-the-ordinary." One he labels a "fuzzy" meteor. "This meteor," he said, "is not characteristic of the majority seen, in deceleration, quality

84

of light envelope, lack of distinctive 'edge' to the path, a very brief afterglow and no color." His other shaggy dog he calls dark meteors, "a brief streak blacker than the sky." He notes that others had seen these along with him and he insists they were not tricks of the eyes.

Eric Kurczewski of New Jersey also believes that "dark" or "nebulous" meteors may be evidence of the demise of small comets. "Unlike a trailed meteor," he said, "a nebulous meteor appears as a blur...and is of very low surface brightness, only slightly above the light level of the background sky. The length of travel of this type of object...is five to ten degrees across the background sky. Very dark sky conditions are required to see this event..." He claims to have seen three such meteors, not associated with any meteor showers, "in 200 hours of observing far from city lights."

The number of references to nebulous meteors was striking. These people were obviously sincere. But I had never heard of nebulous meteors. So I got in touch with Edward Ney, a top-notch astronomer now retired from the University of Minnesota. He had never heard of nebulous meteors either. But I later learned that most professional astronomers pay little attention to the topic. It seems that a century ago in Great Britain the reality of nebulous meteors was a hotly debated topic. There are apparently many solid observations of this phenomenon from amateur astronomers. Nebulous meteors are generally thought to be loose clumps of fine debris of cometary origin. But unfortunately there is no organized information on these observations. There is nothing that allows us to document what is happening here. So I cannot say with any certainty that nebulous meteors have any bearing on the topic of small comets.

Then there was the letter from Keith Bigg, a physicist from Sydney, Australia. He had, he said, "direct evidence," a picture of what might be a small comet entering the atmosphere back in November of 1968. His account had appeared in *Nature* the following year. Bigg had been studying dust layers in the stratosphere with a camera mounted on a high-altitude balloon when the camera happened to snap a picture of a group of meteor trails with a red glow beneath them. The brightness of the trails suggested the object was "very large" before fragmentation. Bigg had been puzzled by "their short paths" and by the lack of "bright flares typical of large meteors." He now believes this could be evidence for our small comets. It is, of course, quite possible for a large object or comet to split in the upper atmosphere and leave all sorts of trails. So Bigg may be right.

A mystery object photographed by a DSMP (Defense Satellite Meteorological Program) satellite in the late 1970s may be yet another portrait of a small comet. One infrared image from this weather satellite shows a strange object above the cloud cover. The object is very white, indicating the "maximum cold" the satellite could register, or less than minus 76°F. Its shape appears almost rectangular but may not be because the DSMP imager does not photograph a whole area at once the way a camera does. Instead, like Dynamics Explorer, it creates a picture one line at a time. So any relative motion between the object and the satellite while these scan lines were collected could have distorted the shape of the object in the image.

Bruce Maccabee, a Maryland physicist who works for the Navy, performed a thorough analysis of the image on his own. He determined that the size of the object could be anywhere from tens of nautical miles, if the object was located near the Earth's surface, to as little as tens of feet, if the object was hovering within several hundred feet of the satellite. Unfortunately, there was no way to determine the object's altitude. But Maccabee showed quite clearly that the object was not created by random electronic or mechanical noise. Nor could it be explained as a cloud or aircraft flying near the Earth itself. And he thought it unlikely that the object was another satellite or piece of space junk traveling in an orbit beneath the DSMP satellite. Maccabee, as director of the Fund for UFO Studies, obviously suspected a UFO.

I estimated the size and speed of a cometary water cloud at a few hundred miles altitude and concluded that the DSMP picture could be of such an object. But in order to prove that this object is a cometary water cloud, other sightings would have to be made with this satellite in order to establish their frequency and hence the number of such objects. Because the speed of the DSMP spacecraft must almost match that of the cometary water cloud such events are expected to be rare, perhaps as infrequent as once every several thousand orbits or more. Other kinds of features, not as spectacular or easily identified, such as short white lines in single or consecutive scan lines could also be expected in the images. What this makes clear, however, is that unidentified phenomena are observed with spacecraft. But I cannot say with any certainty that this particular observation relates to a small comet.

I avoided many topics in the course of the debate on the small comets and UFOs was one of them. Our hypothesis was already controversial enough. We had angered thousands of scientists. We did not need to bring such a taboo subject into the fray. But there is no longer any need to avoid the subject.

A number of people suggested to me that perhaps some of these objects entering the atmosphere could produce the illusion of a flying saucer. Alvin Lawson was one of those who had noticed that the small comets share many ephemeral qualities with UFOs. Lawson, a professor of English at California State University in Long Beach, has investigated UFOs for over a decade and is a skeptic on the issue. He wrote to me saying that if UFO sightings have a cause, then the small comets might be "the most likely explanation yet."

"Might one assume," he asked, "that dust, released gases, and varying sizes and speeds of the comets would result in occasional visibility? I am aware that most of the numerous cometary events occur at high altitudes; but if only one percent of them incinerate there would be potentially 288 'weird lights in the skies' each 24 hours worldwide."

Such an explanation, Lawson believes, could well account for the vast number of UFO reports, as well as their widespread origin, long history, sudden appearances and disappearances, odd behavior, and virtual silence. And if small comets are the cause of UFO sightings, he mused, how ironic it would be: UFOs would actually have turned out to be extraterrestrial.

I think Lawson may be right. In our computer modeling of the comets and how they plunge through the atmosphere, we came across a result that we did not advertise. We learned that water clouds produced by the occasionally larger small comets, those that penetrate more deeply into the atmosphere, would have a peculiar shape. If you look at how these water clouds are deformed, how they might become frozen, then you can see how someone might go out in full daylight and look at the sky and see a saucer-shaped object that looks like it is moving very, very rapidly.

Normally, of course, a small comet should not be visible during the day. And at night you should simply see a short streak across the sky. The average small comet does not penetrate deeply enough into the atmosphere. But not all the comets are expected to be the same size. Some will be large. Of course, the larger the size, the fewer such comets we would expect to see. But also the larger the size, the deeper into the atmosphere they would fall. If the comet is large, but not large enough to hit the Earth, then as it descends into the cold parts of the atmosphere ten to twenty miles above us, the hot steam would slow down and perhaps cool sufficiently to condense into ice crystals and become visible. The underside of the water cloud would be shaped like an ablation cone. It would look just like a curved saucer.

Human perception is such that someone seeing this sort of an object would have considerable difficulty judging whether it was located a hundred yards away or twenty miles away. Then, as the ice crystals on the surface of the cloud disappeared, the object would appear to diminish, and the observer might get the impression that the object was receding at tremendous speeds. This shrinking cloud could then be picked up by the winds, producing the illusion that the object had made a sudden sharp turn. My scenario, of course, will not account for the reports of tiny aliens snatching frightened humans from their homes and taking off from pumpkin patches in flying saucers.

I tread across this territory carefully. Many others have explored this murky, no-man's land of unexplained phenomena before, though most not so cautiously perhaps. One forerunner was a Bronx-born clerk named Charles Fort. In 1919, he published *The Book of the Damned*, his first work of odd data culled from the scientific journals and newspapers of his day. Fort had a fondness for tales of "aerial icebergs" and "shafts of water" from "blue skies," and he poked fun at the efforts of science to explain them. To Fort the reports of falling ice suggested "that there is water—oceans or lakes and ponds, or rivers of it—that there is water away from, and yet not far remote from, this Earth's atmosphere and gravitation.... Sometimes fields of ice pass between the Sun and Earth." Fort almost got it right. I think the small comets may well account for at least some of the many unexplained reports we have of falling ice.

There is a long history of ice chunks of all sizes falling down from the sky, sometimes even crashing through homes. Such tales of falling ice dot the popular press and meteorological journals worldwide. Some of the pieces are said to measure as much as twenty feet in diameter and weigh hundreds of pounds. Though most such hydrometeors, as they are known, are considerably smaller, they are still too large to be explained by the prevailing theory of hail formation in storms. Consequently, hydrometeors are usually blamed on passing aircraft, even though reports of the phenomenon predate the Wright brothers. A study conducted some time ago by the late James McDonald, an atmospheric physicist, found that only two out of thirty ice falls in the United States in the 1950s could be attributed to aircraft.

One of the best documented falls, according to Arthur C. Clarke, who also seems to have an interest in such things, occurred over a tree-lined street on the outskirts of Manchester, England on the 2nd of April 1973. It was just after 8 PM when a Dr. Richard Griffiths saw a large object hit the road just outside a liquor shop. The ice block was quite large, according to Griffiths, who picked it up and stuck it in his freezer overnight. The next day he took the four-and-a-half-pound chunk of ice to his lab at the Manchester Institute of Science and Technology for analysis.

Griffiths tested it the way a hailstone is tested to determine its history. He sliced it into thin sections and found that the ice was made up of fifty-one layers separated by some thinner layers of trapped air bubbles. The regularity of the ice layers and the size of the crystals made it unlike any hailstone he had ever seen. It was also considerably larger than the largest scientifically verified hailstone, the 1.67-pound Coffeyville Hailstone, found in the town of the same name in 1970. Griffiths then made inquiries at the local airport. He discovered that one plane had landed just after the ice block had fallen, but engineers informed him that there had been no icing on the aircraft. Clarke believes that this type of ice fall "ties in beautifully" with my small comet hypothesis.

He is not alone. In 1989 I received a letter from Hu Zhong-Wei, an astronomer at Nanjing University in the People's Republic of China. He told me that on three occasions—the 29th of December 1982, the 11th of April 1983, and the 17th of November 1984—large chunks of "natural" ice had fallen from the sky near the city of Wu-xi. He performed a detailed chemical analysis of a fragment from the latest fall, found that it differed from hail, and concluded that perhaps the ice had come from a comet.

I take the possibility of such falls seriously. If the comets are large enough to begin with, chunks of ice could reach the ground. Under certain conditions, perhaps the center of the cloud does not vaporize at all, or turns to ice in the lower atmosphere, and then crashes down to the Earth's surface. We expect the small comets to vary in size from time to time. Not all of them will be a hundred tons. A few will be smaller, a few will be larger. So for every thousand normal-size small comets that come in, maybe there will be a larger one. Some may be as large as 10,000 tons.

The very largest of these comets are certain to leave their calling cards on the Earth. One certainly was left at Tunguska in Siberia in 1908. What caused the giant explosion there has been a puzzle for some time. But I think it was a comet. It formed a shallow crater and felled trees for several hundred square miles around it. That is exactly what we now expect to happen when a large water comet plows into the Earth. The impact created only a shallow crater, which is not surprising since the impact pressures for a water comet are ten to fifty times less than for a solid rock. But the heat from the incoming hot piston of gas charred the area around the crater. And the atmospheric pressure wave it created left behind an eerie landscape of felled trees, all pointing away from center of the impact.

There may be other Tunguskas around. The flat-floored crater at Flynn Creek, Tennessee, may be one. These events will happen once in every 100 or 1,000 years; for every billion small comets you are going to have a large one that measures hundreds, perhaps thousands of feet in diameter. No one can deny the existence of these larger comets.

If the small comets, these icy interlopers, are real, then there probably exists a good deal more evidence of their presence within all the reports of strange and unexplained phenomena buried in the scientific literature. Oceanographic and meteorological journals have long published reports of bright flashes of light lasting from one to several seconds and covering large portions of the sky. Bright luminous patches have also been seen close to the horizon or moving steadily and deliberately across the sky. And large amounts of water vapor high in the atmosphere might account for a wide variety of optical effects that have been reported, including odd halos and strange rainbows.

The small comets may also be responsible for a phenomenon described by pilots and others who have seen it as "an expanding ball of light." A study of fifteen such cases, taking place between 1959 and 1985, was conducted by psychologist Richard Haines and published in the *Journal of Scientific Exploration*. Pilot reports indicate that the initial ball of light forms at a relatively high altitude. The light, usually white or yellowish, then gradually expands until it becomes increasingly transparent to stars in the background. Eventually it fades completely from sight. Some reports also tell of a towering cumulus-like cloud which appears to rise out of the stratiform layer. I think these clouds might be a comet's vapor trail at lower altitudes. But the phenomenon would have to be related to the atmospheric impacts of somewhat larger comets than those responsible for the atmospheric holes. And we expect that the number of such large comets, penetrating deeply in the atmosphere, to be much less.

Some pilots may also on occasion experience air turbulence caused by the infall of these larger comets. One pilot, Harold Blake of Florida, wrote to me pointing out the existence of a phenomenon called Clear Air Turbulence, or CATs. These, he said, are "elusive and destructive clear air turbulences that most military pilots fear and commercial and private pilots avoid 'like the plague'." They happen without warning. I think that a piston of gas coming down at a hundred miles an hour from one of these comets could cause such turbulence. Near the ground, Blake believes, CATs might also account for wind shear. But there is very little research in this field. Some of these CATs are probably associated with the jet stream in the atmosphere, but are there CATs that are not? To be certain, we would need to see airline records on CATs, if such exist, then calculate their rates and see if they might match a reasonably expected rate of infall of the occasional larger comets.

One pilot who regularly flies the Pacific wrote to me saying he has sometimes seen round holes through clouds. Others have seen them as well. These holes may be as large as several miles in diameter, and appear in otherwise dense, uniform cloud decks. There are also some credible observations of precipitation from cloudless skies. And sometimes layers of unexplained haze are detected by balloon-borne instruments at altitudes above fifteen miles. Could all these be effects due to the small comets?

In 1976 Frank Gibson, of the Air Force Geophysics Laboratory in Bedford, Massachusetts, published a letter in *Nature* entitled "A Rare Event in the Stratosphere." On the 20th of June 1974, a high altitude balloon was sent up in the clear skies over New Mexico to measure the scattering of light by aerosols, or tiny solid particles, in the atmosphere. Upon reaching an altitude of about ten miles, a strong signal was recorded indicating increased aerosol concentrations, a cloud in other words. This haze extended at least another seven miles up and had a reach of some fifty miles across, but it was not visible from the ground. A momentary injection of aerosols would be necessary to form such a cloud. But there was no obvious source for it. Whether it might be related to a small comet is hard to tell.

There is no end to such curiosities. Recently scientists at the University of Illinois claimed to have found clouds of sodium atoms that rapidly appear and disappear at an altitude of about fifty miles. They observed these sodium clouds with an instrument called a lidar. The lidar sends up pulses of laser light into the atmosphere and measures the intensity and wavelength of the light that returns. The scientists, headed by Timothy Beatty, suggest that the clouds are vaporized meteor material. Or perhaps, once again, cometary material. It would certainly complete the recipe for a salty ocean.

Quite obviously there is a whole folklore of flashes and dimmings and strange things going on in the sky. It is tempting to see the small comets as an explanation for many of these puzzling phenomena. But whether these things are related to the small comets is not certain. The phenomena are interesting, but they are often lacking in documentation and so are difficult to interpret. Some could be due to the passage of a small comet, who knows? At the moment, all that we know for certain is just how little we know.

Chapter 12

A Scramble for Satellite Data

Meanwhile, the debate raged on. My discovery of the small comets with Dynamics Explorer had sent a number of scientists off on a frantic search through the data libraries of other satellites. It was a sensible thing to do. After all, hundreds of satellites had been placed in orbit during the past quarter century or so since the launch of Sputnik and probably less than ten percent of all the data received from these spacecraft has ever been closely looked at. So it was reasonable to assume that, if the small comets were real, some other satellite might have already gathered evidence of the passage of these objects through the Earth's atmosphere. Most people were, of course, quite convinced that the reason no satellite had ever reported such objects was simply because the small comets did not exist.

Of those hundreds of satellites, however, only a few were likely to have recorded the signs of the impact of cometary water clouds into the Earth's atmosphere. One of these was the Orbiting Geophysical Observatory known as OGO-4. The Naval Research Laboratory placed an instrument on OGO-4 that provided continuous observations of ultraviolet light for a period of about eighteen months during 1967-1968. Nearly twenty years later Robert Meier and Talbot Chubb, physicists at the Naval Research Laboratory's E.O. Hulburt Center for Space Research, located in Washington, D.C., decided to reexamine the measurements from 2,000 of the satellite's orbits for evidence of atmospheric holes. Shortly afterwards they submitted a paper to *Geophysical*

Research Letters saying they had failed to find any such evidence.

I refereed the paper at Dessler's request. The paper had one glaring problem. Their data did not support their conclusion. The other referee picked up on this as well. There was an abundance of unidentified fluctuations in their dayglow observations that might be ascribed to the presence of atmospheric holes. A casual examination of even the very few orbits in the published literature revealed a number of possible events. In fact, there appeared to be one or two events in every orbit. In a revised version of their paper, Meier and Chubb presented all kinds of excuses for the dark spots in their measurements. I think they went a little out of their way to avoid the possibility of atmospheric holes.

They claimed that a draftsperson had added a few bumps which were not actually present in the data. Other blips, they said, were caused by sunlight, tropical arcs, particle precipitation, periods of lost spacecraft orientation, and auroras. You name it and it was there. It was clear that the authors had decided from the start that all the fluctuations in their observations were due to something other than cometary water clouds, even though no definitive identification of every isolated dark spot was possible. In fact, very few of the events in their dayglow observations were likely to be caused by cometary water clouds—just one in every 100 to 1,000 orbits. But the authors had failed to convincingly identify each dark spot in their data, so they could not convincingly exclude the presence of cometary water clouds with the OGO-4 observations. Their paper was rejected.

Our Dynamics Explorer spacecraft had a twin called Dynamics Explorer 2. Both were carried aloft, piggyback-style, on the same rocket back in 1981. Scientists hoped that this pair of satellites would give them a ringside seat on the Earth's royal crowns, its auroral lights. We wanted the big picture, in other words, not just the worm's eye view aurora chasers get from down here on Earth. Our Dynamics Explorer was stationed at a high altitude to study how this natural wonder is produced, how a beam of electrons from space hits the Earth's atmosphere as if it were a giant television screen and produces these auroral "pictures." Dynamics Explorer 2 was supposed to cover the same territory but from a lower altitude, skimming over the upper atmosphere, just above the "screen."

Every few months there were working group meetings on the Dynamics Explorer mission. They were usually held at the Goddard Space Flight Center in Greenbelt, Maryland, and each one lasted a couple of days. From time to time I would present our findings on the atmospheric holes at these working group meetings. Some people not involved in the Dynamics Explorer mission would even come in and listen. It created considerable interest. One of the people who was kept up to date on our findings and interpretation of the holes was Bill Hanson, the director of the Center for Space Sciences at the University of Texas in Dallas. I took great pains to let him know exactly what we were doing on the atmospheric holes, but he got very emotional about it and was to remain so throughout the controversy.

Hanson had an instrument aboard Dynamics Explorer 2. His instrument examined the contents of the ionosphere, the Earth's radio-reflective covering that consists of ionized gases. It is here that the Sun's ultraviolet radiation tears neutral gases in our atmosphere apart, producing ions and electrons. The result is a veritable shroud of ionization, or charged particles, around the Earth. Hanson was certain that if the small comets were real they would be causing drastic effects to the Earth's ionosphere. So he speculated on how the ionosphere would be affected by the passage of a cometary water cloud. He believed that the water vapor from an impacting comet would cause a mammoth hole in the ionosphere that would only return to normal about ten minutes after the passage of a small comet. He thought that a satellite moving through the ionosphere, such as Dynamics Explorer 2, should encounter such holes about once in every two hours of satellite flight.

Unfortunately Dynamics Explorer 2 did not have a long life. But it did manage to collect data for a year and a half before falling out of its orbit. Hanson ended up examining only five hours of measurements from Dynamics Explorer 2 and another eighty hours of data from two older spacecraft, Atmospheric

Explorer C and Atmospheric Explorer E. Hanson, however, found absolutely no unusual perturbations of the ionosphere. He concluded, in his Comment published in *Geophysical Research Letters*, that either we had overestimated the number of small comets impacting the Earth's atmosphere or that the ionospheric effects of these objects were "more subtle" than he had assumed.

We also examined Hanson's Dynamics Explorer 2 data. But unlike Hanson's, our search did reveal some evidence of the passage of small comet water clouds through the ionosphere. We found it by looking first at those times when the Dynamics Explorer 2 had passed near an atmospheric hole that had been spotted by our Dynamics Explorer. We expected near simultaneous sightings by both spacecraft to be rare and in fact we ended up eliminating all but one or two of our 26,900 candidate atmospheric holes.

One of these atmospheric holes occurred near a disturbance in the ionosphere seen by Dynamics Explorer 2. The event happened at low altitudes and was quite brief. This suggested to us that the interaction taking place between a cometary water cloud and the ionosphere was not a large one. Rather than a severe loss of the number of ionospheric ions, the comet clouds appear only to create a small region of turbulence in the ionosphere. Actually, the passage of the cometary water cloud through this region resembles the disturbance caused by a large, fast object, such as the space shuttle.

We then went back and searched for the presence of other similar disturbances in a limited portion of ionospheric data from Dynamics Explorer 2 and found three more such events. Of course, these disturbances may or may not be due to small comets. But their size and frequency are about what you would expect for water vapor clouds produced by small comets as they come plummeting through the ionosphere. That is what we published in our Reply. Hanson had looked into the basket and did not see any apples. When we looked, we saw apples. Hanson seemed taken aback and angry. I think we had taken him by surprise.

On the 22nd of February, 1986, just six weeks before news of the small comets first hit the newspapers, the Swedes launched a satellite called Viking. Like our Dynamics Explorer, their Viking satellite was equipped with an ultraviolet camera designed to take pictures of the aurora. It had two imagers actually, both provided by the Canadian government. This made Viking, of all the satellites in the sky, the most likely candidate to confirm or deny the existence of the cometary water clouds above the atmosphere that had first been seen by our Dynamics Explorer.

So, not long after Viking began sending back its images of the Earth's royal crowns, the press began to inquire about whether or not the satellite had seen any "atmospheric holes." The job of fielding these questions fell to Sandy Murphree, a physicist at the University of Calgary who was then the principal investigator for the Canadian imagers. Murphree, it so happens, had obtained his degree at Rice University while Dessler was chairman of the Space Physics and Astronomy Department. It was a small world.

The situation proved to be tremendously distracting for Murphree and his small team of Viking scientists. Few were asking questions about their beautiful pictures of the aurora and this, I am sure, they found distressing. I never realized that the publication of our original papers would cause such a problem for them. If the roles had been reversed, I also would have been tremendously annoyed. Besides, scientists often spend years studying the data from their satellite instruments. The question about atmospheric holes was not an easy one to answer. You have to look very carefully at your images. You really have to do your homework. So I understand how the Viking people felt.

Unfortunately, the press wanted a quick yes or no answer. But scientists and journalists usually march to the beat of different drummers. Science is slow and cautious. The press is always in a rush and not always cautious. They forced Murphree and his Viking team to issue a quick statement. And what quick statement could they make other than to say, "We don't see dark spots in our images"? Of course, they had not done an extensive search for these holes and never claimed that they had.

Early in 1987 the Calgary team published a paper on their Viking imager, which briefly mentioned this negative result. They had inspected a "small sample" of the images but had found no cometary holes of the sort we had reported with Dynamics Explorer. A more "extensive analysis" would be needed, they said, to confirm or deny the existence of the small comets. Once again, Dessler had asked me to referee the paper for *Geophysical Research Letters*.

I suggested a small qualification. I thought the investigators should clarify whether their instrument could have picked up such holes, if they did exist, and under what conditions it could do so. I wanted the Calgary team to say that they had not completed the analysis of their search for the atmospheric holes, and that by just looking at the pictures they could not tell if a dark spot in their images was an atmospheric hole or not. Dessler told me that he would take care of it, but he never did. This precipitated a very bad situation. Everyone got the impression that the Viking people had searched their images very carefully. But you cannot do a thorough search in such a short time and be sure it is correct.

Sandy Murphree and his team said they had looked for, but not found, any trace of the small comets in their images. But when we first saw their images during a presentation of their findings at a meeting of the American Geophysical Union in the spring of 1987, we could see things that looked suspiciously like atmospheric holes. They did not appear as frequently in the Viking data because the satellite was in a lower orbit than Dynamics Explorer. This lower vantage point offered Viking a smaller view of the Earth's atmosphere and consequently fewer small comet atmospheric holes would appear in their images. Besides, the Viking had a different imager on board than the Dynamics Explorer and therefore a different analysis was required.

We suggested a collaboration. We had already agreed to work together on the aurora. The beauty of the Viking measurements of the aurora is that they could look at the North Pole while Dynamics Explorer looked at the South Pole. It was a fantastic opportunity to find out whether the aurora looked the same over both poles. We were not about to sacrifice this sort of opportunity over a petty argument over atmospheric holes. But they agreed to collaborate on them as well.

So in January of 1988, Sigwarth, Craven and I travelled up to Calgary to start our collaboration on the aurora and examine certain Viking images for the possible presence of atmospheric holes. They had two cameras aboard Viking. Unfortunately, the camera that provided them with the most images was not suited for spotting atmospheric holes. It had a noisy sensor and the camera appeared to be picking up stray light reflected from atmospheric clouds. But it was still well suited for auroral work.

The other Viking imager had a coated mirror that provided protection against cloud reflections and we saw no reason why it would not be able to detect holes in the atmosphere produced by the small comets. Unfortunately, the camera had died on the 5th of May, just a couple of months after launch, so there were not many images. In addition, most of the images from the failed camera had been taken from the Earth's nightside to see the aurora, rather than from the dayside, where atmospheric holes should be visible.

So we began to examine the few existing dayside pictures. Murphree was not quite sure what atmospheric holes would look like in his images. He pointed to some dark spots and asked us if they were blackened due to some electronic malfunction. Then we changed the color scales on the television set to get better contrast and the spots began to look very much like the holes in the Dynamics Explorer images. It was immediately apparent to everybody—including Murphree—that if you looked carefully at the pictures from the Viking camera that should see the holes, you saw things that indeed *looked* like holes. Of that there was no doubt.

So in May of 1988 we presented a joint paper at the American Geophysical Union meeting in Baltimore. I gave the talk in Murphree's absence. I had told him what I was going to say. I would show that there were holes in the Viking images. The Viking images even showed the atmospheric holes in greater detail than in the Dynamics Explorer images. The satellite was at a lower altitude and therefore closer to the holes and its camera had a better resolution than ours had. So the Viking images showed these holes not in one or two pixels, or picture elements, but in clusters of six or seven. They were globs of pixels. Yet the holes were the same size as those seen by Dynamics Explorer. They were also of the same blackness and, even more importantly, they occurred at the same rate. The probability that a half dozen pixels could randomly come together in an image to form such a dark spot was extremely small. I knew the results would cause a furor.

Reporters quickly called Murphree. It is unfortunate that this debate has been so public. You could not sneeze without having your handkerchief exhibited for all to see in the newspapers. Afterwards the press reported that the Calgary group was withdrawing from the paper. But that was not true. We agreed to work side by side until a mutually acceptable version of the paper could be found. No one realized at the time that the task would take a year and a half. And I did not realize that this cooperation eventually would fail.

Problem number one was that the Viking images were noisier than those from Dynamics Explorer. In particular they have an amplifier noise that produces diagonal lines across the image. So when you do a detailed investigation, it is very hard to show convincingly that the atmospheric holes could not be due to noise. We first examined the laboratory calibrations of the camera and then decided that an even more rigorous in-orbit calibration would be necessary. If the holes were noise they should appear randomly throughout an image. If the holes were real, they should only appear where we expected them to appear.

Problem number two was that the Viking imager had produced few usable pictures for such analysis. But there seemed to be just enough images to get a satisfying check on the camera to see if the holes were real or just artifacts. To understand how the calibration works you must realize, first of all, that most of the glow appearing in these ultraviolet images is produced by atomic oxygen when the camera is looking nearly straight down into the atmosphere. This is called nadir viewing.

But it is a property of the atmosphere that as you look further and further toward the edge of the Earth, which is called limb viewing, there is a greater intensity coming from nitrogen molecules. Since the water vapor from the small comets does not absorb the light from nitrogen very well, you should not often see holes near the limb. We found this to be true in the Dynamics Explorer pictures. When we examined the Viking images we found seven dark events when nadir viewing and only one for limb viewing in the same set of images. This was a profound confirmation that the darkened spots in the images were indeed not noise but actual atmospheric holes.

We exhausted all possible avenues for analysis. It took a

tremendous amount of work, both on our part and by the Calgary group. But in the end the Calgary group was not completely convinced that the holes were not noise, although they were never able to identify a specific electronics failure or malfunction that could have produced such an effect. Perhaps they still felt bound by their previous hasty statement. In any case, they would not reverse themselves and in June of 1989 the Calgary group finally decided to withdraw their names from the paper. There was no bitterness between us. But we were disappointed. We felt that the findings of Dynamics Explorer had been confirmed.

One last feature of the Viking images is worth mentioning. Next to the darkened pixels representing the atmospheric holes in the images were adjacent pixels that showed a bright light. The images from Dynamics Explorer sometimes showed a bright region around an atmospheric hole but most often this brightening was not seen, probably because of the camera's poor resolution. But the Viking camera had a better resolution than Dynamics Explorer's. If the dark spots in the images really were holes, then bright rims should appear around their edges as the clouds of water vapor came crashing through the Earth's atmosphere. The face of the cometary water cloud should be hot due to the collision with the thin atmosphere at these altitudes. The Viking images showed these bright rims on its atmospheric holes.

The Viking scientists, which included the Calgary team, had working group meetings just as we did on Dynamics Explorer. Apparently some people at these meetings would point to those bright rims around the dark holes and ask what they were. They are noise, the Calgary people replied. Murphree had faced the same problem we had. When we first presented our Dynamics Explorer images at working group meetings, people would see the pretty aurora and ask, but what are those black spots? The Viking scientists had a similar question for the Calgary group: "What are those bagels?" they asked. Everybody referred to the black spots with bright rims as bagels.

Chapter 13

Extraordinary Evidence

Extraordinary claims excite the media, but their staying power is often fleeting. The recent flurry of excitement over cold fusion is a good example. The media trumpeted the claim in its usual way—as a fantastic new discovery. They had done the same with the small comets. But there was a difference. The chemists who announced the startling results of their cold fusion experiments did so without having firm evidence in hand. There was no published paper. (An informal paper was circulated privately and I received my first copy from a Los Angeles stockbroker who was once a student of mine.) Those who tried to reproduce the cold fusion experiments failed.

Our claim also was extraordinary. But our results had been published in a pair of refereed papers and the Viking spacecraft had, in effect, confirmed our findings with Dynamics Explorer. The evidence we had on hand was nearly overwhelming: tens of thousands of pictures showing the holes in the atmosphere caused by a previously undetected population of small comets. Each one of these pictures was worth much more than the proverbial thousand words. They are worth more than ten thousand words, actually, since each standard image is composed of 18,000 pixels.

Each of these pixels, or picture elements, has a lot to tell us about what we are looking at. But like words, pixels can sometimes mean different things to different people. Some of our critics interpreted the darkened pixels in our images as noise in our instrument rather than as clouds of cometary water vapor in the atmosphere. But the simple fact was that instrumental errors, telemetry problems, and computer malfunctions occurred far too infrequently to account for atmospheric holes. Our opponents challenged us on this point repeatedly during the year and a half that the debate took place in *Geophysical Research Letters* and we were forced to tread this ground over and over again. But rather than simply deny their claims, we utilized these opportunities to publish an in-depth analysis of the properties of these atmospheric holes. What we had found confirmed that the phenomenon was indeed very real.

Our opponents used several strategies to blame the appearance of atmospheric holes on instrumental noise. One concerned the size of the holes with respect to the satellite's altitude. If these holes were for real, they argued, then they should appear much larger—and consequently cover many more pixels in the images—when the satellite was at its closest approach to the Earth. Talbot Chubb of the Naval Research Laboratory in Washington, D.C., was the first of many to make this argument. At its closest point, or perigee, Dynamics Explorer flew at about 350 miles above the Earth. At its farthest, or apogee, the satellite orbited at an altitude of about 14,350 miles. Chubb claimed that the holes were not any larger at the lower altitudes. We found that they were. And we had the pictures to prove it.

To anyone with a clear understanding of how our satellite acquires its images, these pictures and our Reply to Chubb were devastating. Our ultraviolet camera works much like a FAX machine in that it produces its images line-by-line. The camera obtains its lines one picture element at a time, pixel by pixel, in other words. Each pixel is a measure of the light intensity of a small portion of the Earth's dayglow and is recorded during 3.9 thousandths of a second by one of the camera's two count registers. The registers alternate in producing the pixels in each line. A line of pixels is recorded as the satellite makes one rotation, which takes six seconds. After each rotation, the scanning mirror which feeds the light intensities into the count registers is tilted another "step" to obtain the next line of the image. Each image consists of 120 lines of pixels and takes twelve minutes to transmit to the ground.

A careful examination of our images showed quite clearly that the majority of atmospheric holes were no larger than a single pixel when the satellite was at high altitudes. Only a few holes were two pixels wide at high altitudes and when this was the case the additional pixel occurred *in the next scan line* and next to the first dark pixel. It was as if we were tracking the movement of the cloud from one scan line to the next, like pointing a rifle at a distant moving object. This situation occurred ten times more often than what you would expect by chance. Sometimes the cloud could even be traced for three or four scan lines. This is just what you would expect if the water vapor clouds were far away and had an apparent speed of about 20,000 mph.

We also found two occasions in which the holes blackened three consecutive pixels *in the same scan line*. Both times the satellite was at low altitudes. These events occurred at a rate that was about forty times greater than chance. But this was the rate expected for such events, because the probability of sighting an atmospheric hole diminishes very rapidly at lower altitudes where a much smaller patch of the atmosphere is viewed than at higher altitudes. The rate of detection was also diminished at low altitudes by the fact that the spacecraft itself spent very little time there. Yet when a cometary cloud was spotted while the satellite was low, it appeared larger than when the spacecraft was high. This is exactly what you would expect to happen. So instead of appearing as a single pixel in a scan line as it did when the satellite was high, when the satellite was low, the cloud was seen as a string of dark pixels in a single scan line of the image.

But the clincher was this. A satellite at low altitudes is closer to the cometary cloud than at high altitudes. So the cloud would move too rapidly to appear in the next scan line when the satellite was low. By the time the camera records the next scan line six seconds later, the cloud would be far beyond the view of the camera. That is just what happened. We have no images of holes appearing next to one another in adjacent scan lines while the satellite was at low altitudes.

So Chubb's attempt to dismiss the small comets on the basis of an altitude dependence failed. But he covered himself with other criticisms. He argued that even though our instrument sampled the dayglow sequentially by two counters, both counters shared some of the same electronics. A failure in one of these shared

parts, caused perhaps by a cosmic ray, could create an atmospheric hole, said Chubb. But we had long ago considered this
possibility and rejected it. Our instrument was relatively im-
mune to cosmic rays.

None of these criticisms upset me. But I was angered when
Chubb said we had treated some of the atmospheric hole images
in such a way as to make the holes look larger than they actually
were. Not true—all the images had been processed the same
way—the way the images from many satellites are processed, in
fact. Chubb had also consistently misspelled Sigwarth's name in
his Comment, putting an "o" in place of the "a."

I wrote to Dessler asking that these items be corrected. They
were not matters of interpretation. They were straight factual
errors. Dessler should have forwarded my comments to Chubb,
but I do not know if he ever did. Perhaps I should have
communicated with Chubb directly. I am sure that he would
have corrected the errors. In any case, Chubb's Comment, errors
and all, was published in the October 1986 issue of *Geophysical
Research Letters*. It was very frustrating.

Bill Hanson was also convinced that our instrument was in
error and that the barrage of hundred-ton comets hitting the
Earth was a fantasy we had conceived. Earlier, he tried to show
that a large population of small comets was inconsistent with the
results he had obtained with our sister satellite, Dynamics
Explorer 2. Then, just a couple of months later, he seemed to be
after us again. In June he asked to have some of our images. He
did not tell us what he was going to do with them. But we very
promptly sent the magnetic tapes for the images down to him at
the University of Texas at Dallas.

A few weeks later one of Hanson's colleagues, Bruce Cragin,
showed up at the Holiday Inn in Iowa City for a meeting on
another subject, Jupiter. But he had brought with him a paper
saying that the atmospheric holes in our images were instrumental noise. I like Cragin. He is a good physicist, but he is heavily
influenced by Hanson. He handed me the paper in my office and
I could tell he felt a little nervous. I looked through it and saw
that it dealt with an old problem.

We had spent three years trying to find a way for the instrument to take the blame for the black spots in our images and failed. Hanson, who was first author, along with Cragin and two of their colleagues, worked on the problem for just a few weeks. They found that "the dark pixels are primarily single pixel events at all altitudes" and concluded that the atmospheric holes were an instrumental artifact. This, despite their admission that they had been "unable to identify a design problem or possible failure mode in the counting circuit that could account for the dark pixels."

Dessler rejected their Comment on the grounds that Talbot Chubb had brought out essentially the same point earlier. But Hanson, like the others, was out to get us. He had lost the earlier round and now he was striking back. He seemed totally consumed by his efforts to find something wrong with our work. So he and his colleagues revised their paper. This time Dessler accepted it. I never understood why Hanson was allowed to publish two Comments. The usual ground rule is one Comment from any single individual or group. But then again, Hanson and Dessler are good friends. They had been at Lockheed together and then at the University of Texas at Dallas.

Their revised Comment was thinly disguised. It listed Cragin as first author and Hanson as one of the three other authors. The conclusion was much the same and so was much of the material. They still harped on what they perceived to be our altitude problem. They claimed that the frequency at which the black spots was observed remained the same regardless of the spacecraft's altitude. But if you look carefully at the results in their paper, the frequency of the atmospheric holes *does* vary with altitude. I really think they wrote the conclusion before they even saw the data. I sent them a letter directly pointing out that they had found an altitude dependence. But they ignored my letter.

Their argument did have one new thrust, however. They were aware that the pixels that comprised each image overlapped somewhat. So, they argued, if the camera is taking a string of pixels and one of them is dark due to an object of some kind, then the pixels on either side should show some signs of darkening as well. They were, of course, quite correct. They calculated that these adjacent pixels should be darkened by about nine percent but found that they were not darkened beyond about one percent. Since there was not enough darkening of the adjacent pixels, they concluded that the holes were simply noise.

Though their argument was essentially correct and presented with great force and conviction, their arithmetic and interpretation of the images were all wrong. When we did the same analysis, we found that there was in fact some darkening of the adjacent pixels. But it was not the same amount of darkening Cragin and Hanson had expected to find. It was not the amount of darkening expected if the cometary cloud was all dark. Granted, if the cloud was all dark, then the pixels did not fit. But Cragin and Hanson had forgotten that an object slamming into even the thinnest of atmospheres would also have to glow.

A careful examination of the pictures from Dynamics Explorer shows that the clouds of cometary water vapor are not merely dark. The black spots are sometimes accompanied by bright features. When such bright features appeared around the dark spots in the Viking images, they had been dubbed "bagels." Even though the resolution on our imager did not compare with Viking's, the bright features were still visible in some Dynamics Explorer images. We had mentioned these bright features at several of our talks. These cometary clouds would have to glow a little as they plowed through the upper atmosphere at twenty times the speed of sound. But Cragin and Hanson failed to take this feature into account and had assumed that the holes were strictly dark objects.

Their model did not accurately describe what was seen in the pictures. Obviously, if the cloud also has a bright feature to it, the darkening of the adjacent pixels would not be as large as Cragin and Hanson had assumed. We showed that if you took these bright edges into account, you would get exactly the amount of darkening in the adjacent pixels that you see in our images. Rather than trying to impose our conclusions on the images, we let the images draw the conclusions for us. And in this case it was obvious that the instrument had provided substantial evidence for the observation of small, opaque objects that glowed along the edges that took the impact of the atmospheric gases.

We had begun the computer programs involved to reach this conclusion long before we had received the Cragin Comment. Sigwarth had been working on it as part of his master's thesis. Much of the work that he did went into our Reply to Cragin. It would be, at four pages, one of our longest replies, and for Sigwarth one of the most challenging. The problem was straight-forward, but it required the use of many images and much computer time.

We knew there was a luminous edge to these infalling clouds of cometary water vapor but we did not know their shape or form. We first assumed that each dark cloud would appear to be disk-shaped and surrounded by a bright ring. But such an object did not fit the actual darkening of the adjacent pixels in our images. The model that best fit the observed holes was of a dark disk with a bright half-limb. Sigwarth's thesis went into detail on how any change in the shape, brightness, or thickness of this limb would not produce what we were seeing in our images.

Sigwarth did a fine job. He worked out the problem correctly. He did it rigorously and without approximation. It was a time-consuming task that often ran into the wee hours of the morning. But by looking at enough objects, he was eventually able to build up a clearer picture of them. Sigwarth showed that the side of the water cloud ramming into the Earth's atmosphere would have a bright face, while the side facing the camera would be mostly dark with only a little light showing at its edge. The bright face of the cloud could either be due to the high-speed collision itself, or by a handshaking exchange of charges between the cometary water molecules and the oxygen ions in the upper atmosphere.

But Sigwarth's thesis went one step further. He made a prediction. If the bright spots are there and manage to show up against the relatively bright atmospheric screen when the camera is pointing Earthward, then when the camera is looking above the limb of the Earth into the darkness of the sky we should be able to see the bright face of these incoming cometary water clouds. The cloud's dark side would not be seen above the Earth's limb, because the sky itself is too dark. But the bright leading edge of the clouds should be visible as they come burning in through the atmosphere. So he went searching for these bright spots and he found them. The images are stunning.

The master's thesis by Sigwarth is a beautiful piece of work. It is the equivalent of a Ph.D. thesis. When he presented it at the American Geophysical Union meeting in San Francisco in December of 1988, there was standing room only. Despite all the negative comments, our small comet presentations were still drawing SRO crowds. And sometimes even standing room was hard to find. I was sitting a little way behind Hanson and he had a large grin on his face before the presentation began. He was unaware that I was sitting two rows behind him. But a minute into Sigwarth's talk, not even Hanson was grinning. It was tremendous. A lot of people came up afterwards to congratulate Sigwarth on a very impressive piece of work. I think Bill Hanson was very unhappy. I do not know if he had ever lost a public argument before. But here he had lost twice.

Chapter 14

Launch of the Artificial Comets

Not everyone was out to get me. Michael Mendillo was one of those rare other individuals who would conduct an experiment to shed some light on the small comet mystery. Mendillo is a professor of astronomy at Boston University and a space scientist for its Center for Space Physics. His work would show that our proposal was, at the very least, quite plausible.

On September 10, 1988, a Terrier/Black Brant sounding rocket carrying a special payload of explosive chemicals was launched from the NASA Wallops Island Flight Facility in Virginia, 150 miles from the nation's capital. Shortly after noon, the payload was detonated by ground command, releasing 584 pounds of water vapor, ice, carbon dioxide and debris 185 miles up into the atmosphere. Almost four thousand miles above, Dynamics Explorer was watching. The satellite, which had been re-programmed to take a series of one-minute images of the narrow region surrounding the release of the rocket's payload, managed to spot the event.

This was the third and final launch in a series of experiments that Mendillo had designed. The experiments were named ERIC (Environmental Reactions Induced by Comets), after his son. They were meant to simulate the deposition of a cloud of cometary water vapor in the atmosphere. It could not be done exactly, but you could release a water-producing bomb in the upper atmosphere, explode it, and see if you could pick up the event with the camera on a satellite.

You could not, of course, put as much water up there as we

thought the comets were dropping down on us. You could take up a few hundred pounds, but versus a comet's one hundred tons, that was not much. Besides, Mendillo was simply depositing the water cloud at an altitude of less than 200 miles. The water clouds from the small comets pass by this altitude at great speeds and come to rest at an altitude of about thirty-five miles. But despite such differences, Mendillo's cloud should have produced the same kind of effect on the upper atmosphere as a small comet, and it did.

We first met with Michael Mendillo in San Francisco in December of 1983. John Sigwarth had presented our second talk on the mysterious black spots. Some people were starting to get excited about the subject. Mendillo was one of them. He approached us and suggested an experiment that might confirm our suspicion that the black spots in our images actually represented water vapor. What if you fired a rocket that would release a payload of water into the atmosphere? he wondered. How would the imager on Dynamics Explorer record such an event? Would it show up as a black spot also? It was an excellent suggestion. We were interested. But at the time we were too caught up in other projects to devote ourselves to it.

We were to meet again soon. Mendillo and I both had experiments aboard one of the last flights of the space shuttle Challenger, the one that carried up Spacelab 2 at the end of July 1985. Mendillo had an interest in the effects of water on the upper atmosphere and on this flight he and his co-workers used a ground-based radar to track holes in the ionosphere created by the water vapor exhaust from the space shuttle's engines. I knew then that he was determined to carry on his little experiment with us.

Mendillo is strong, ambitious and energetic. He is also a good physicist and conservative. I have a certain requirement in people. You have to be conservative. You have to know physics. And you have to work hard. Mendillo is one of those people. He wanted to know how small an amount of comet-like material would be detectable in the Earth's atmosphere and no one was about to stop him.

It never ceased to amaze me how quickly developments on the

small comets front would spread. The nature of the development, be it positive, negative or neutral, mattered not. Every time anyone had anything to do with small comets, everyone seemed to know about it. Mendillo, with a smile on his face, told me about a strange little episode that occurred shortly after he had decided to go ahead with ERIC. During a conference someone had come up to him and began ranting and raving while he was standing at a urinal in the men's room. "Can you imagine believing enough in the small comets to go out and do something like this?" the man had told Mendillo.

But NASA was receptive to ERIC and funded Mendillo on three separate shots. All three were launched from NASA's Wallops Island Flight Facility. The first took place on the 7th of August 1987, and deposited 300 pounds of water, carbon dioxide, ice, and other gases high up in the atmosphere. Two satellites had been positioned to observe the event. One was the Dynamics Explorer satellite. The other was Polar BEAR (Polar Beacon Experiment and Auroral Research). Polar BEAR was orbiting at a lower altitude than our satellite, so we expected to get a better signature with it. But this little satellite was full of surprises.

Polar BEAR has had an unusual history. It was originally part of a series of navigation satellites for the U.S. Navy. Built by the Applied Physics Laboratory at Johns Hopkins University, these satellites proved to be so reliable and unlikely to need replacement that a "spare" was donated to the Smithsonian Institution's National Air and Space Museum in 1976. It spent the next eight years on display in the museum's Gallery of Satellites.

But in 1984 it was called back into service. The Applied Physics Laboratory made this unusual request after learning that it would no longer have Navy and Air Force satellites available for modifications. The museum was offered, and gladly accepted in exchange, a similar satellite of an even earlier design. The old navigation satellite was then refitted to monitor auroras, magnetic-field changes, and other ionospheric effects due to disturbances such as solar flares. It was placed back into service in November of 1986.

Nearly a year later, there was a failure aboard the satellite—just one day before Mendillo's first ERIC experiment. The mirror on the satellite's ultraviolet camera locked up. It could not scan the entire area in the atmosphere where the artificial comet was released. All it produced was a single line across the picture.

Mendillo and Sigwarth had worked very hard to find the right time to launch the rocket. They had to take into account not only the Earth's rotation but also the orbits and motions of the two satellites. A tremendous amount of coordination was necessary. The pictures had to be taken on the fly. Dynamics Explorer had to be as close to overhead as possible. From its high altitude position, the satellite had no chance of picking up such a small event if an entire picture was taken. So in order to get as many images of the water release as possible during the few minutes it was up there, Sigwarth had to limit the scan of the mirror used by the Dynamics Explorer's camera. The cameras on both satellites were programmed to scan over the section of the east coast of the United States where the water would be released. The launch and water release had to be timed perfectly. Everything had to be done just right. It was a fabulous job. I really had little to do with it, other than biting my nails over how much they were limiting the viewing area of the camera. They could so easily have missed the target area entirely.

We saw a little signature with our satellite on the first ERIC shot and so did Polar BEAR, miraculously enough. There was not a lot of water in the payload so we expected the signature would be small, and indeed the black spot that resulted was quite small. Unfortunately, this decrease we observed in the ultraviolet light was within the limits of normal variability. So perhaps that little black spot was due to chance, but then again, perhaps not. The results were tantalizing.

With Polar BEAR out of the action, Mendillo searched for a backup for Dynamics Explorer before the second ERIC launch. He decided to use the Millstone Hill radar operated by the MIT Haystack Observatory in Westford, Massachusetts. Mendillo thought that this radar installation, operated by John Holt and David Tetenbaum during the payload's release, might be able to track the disturbance caused by an influx of comet-like matter in the ionosphere.

The second ERIC launch deposited some 430 pounds of water vapor and carbon dioxide about 180 miles up in the atmosphere.

Again it was observed by our satellite and again the signature was a small one. But the cloud also produced atmospheric effects that were picked up by the Millstone radar. Within minutes of the release, the cloud caused a well-defined ionospheric depletion, or hole, that was monitored by the radar for about one hour. The location of this ionospheric hole coincided with the coordinates of the darkened spots on the Dynamics Explorer images.

I think Sigwarth and Mendillo were lucky. Each time the rocket drifted off one way or the other, almost enough for the camera to miss the region of the payload's release entirely. Each time I told Sigwarth to allow a little leeway and each time he said: "Oh yes." The first signature appeared at the edge of the image. Before the second launch I asked Sigwarth if he had put in more pad. "Oh, yes," he assured me. But that made me a little nervous. The second shot, on the 29th of January 1988, produced a darkened spot right at the edge of the picture. I asked Sigwarth if on the next attempt he would make the picture a little wider. Again we were lucky. The third shot, which took place the following September, also produced a weak dark spot in the images. But again the Millstone radar observed a well-formed decrease in the ionospheric density that coincided with the location of the darkened pixels on the Dynamics Explorer images.

Project ERIC produced good, if not spectacular, results. We observed the faint dark spots produced with all three injections of water into the atmosphere. And Mendillo found that if you scaled up the size of the atmospheric holes, or the darkening produced by these artificial comets, to those produced by the real comets, the amount of water required would be about a hundred tons. So the weak signature was about all we could expect with these small payloads.

This experiment did not prove our hypothesis. It was a calibration. It was enough to establish the fact that if you have water vapor up there, you are going to have a black spot on the satellite images. The radar results were also encouraging. Mendillo hoped to get an indication of how to take a radar and look in the ionosphere and get a unique signature for cometary material. He now thinks he might be able to use such a radar to pick up the intrusion of small comets into the atmosphere. But I believe the dynamics of the two situations are quite different. Mendillo was just laying water vapor up in the atmosphere, but there is no way of stopping the water vapor clouds injected at high speed by the small comets at those altitudes. Only a trace of the cometary water will be left at these altitudes.

It really did not matter, however, because no one paid any attention to the results. No one cared that during the third ERIC experiment, the camera aboard the Dynamics Explorer satellite captured the signature of one of the real water clouds from the small comets, just some fifty miles away from the area of the darkened pixels produced by Mendillo's artificial comet. Reality, it seems, had reserved itself the right to have the last laugh.

Chapter 15

Some Cometary Competition

I am getting a little ahead of my story because so many things happened at the same time. Just as the debate taking place in *Geophysical Research Letters* was coming to an end, Tom Donahue, like a jack-in-the-box, popped up with a surprise for everyone in the summer of 1987. I was in Vancouver at the time presenting a paper on the Earth's auroral lights at an international conference of space scientists. It was a nice sunny day and I was walking across the lawn going from one meeting to another. Suddenly someone rushed up to me. "Did you hear about the unscheduled paper Tom Donahue just presented?" I had not. "Everyone has been looking for you."

Donahue, I was told, now claimed to have found a new population of baby comets in interplanetary space. I was flabbergasted. This is the same Tom Donahue who spent more than a year trying to destroy the small comet hypothesis. Donahue had been fiercely critical of the small comets even before our original papers had been submitted for publication. Then, when he later refereed these papers, he recommended that they be rejected. And once the papers were published, he was also the first to submit a formal criticism of the small comets for publication. In fact, Donahue had spent more time than anyone else trying to debunk the small comets. And now, to everyone's great surprise, he had discovered a population of such objects on his own.

That was the good news. The bad news was that Donahue had said that his newly discovered objects had nothing to do with my small comets. I was taken aback. How could this be? This was all

117

the more confusing because the data Donahue was now using as evidence for his new population of celestial objects, he had used just a few months earlier to show that no large population of small comets could possibly exist in interplanetary space.

Donahue fired this last shot in our direction in December of 1986. He argued, in a paper submitted to *Geophysical Research Letters*, that if the small comets were as numerous as I had proposed, no amount of protective mantle would be sufficient to halt the vaporization of the comets and prevent the flooding of the solar system with water. The action of sunlight on these watery comets, Donahue insisted, would release large amounts of hydrogen atoms in interplanetary space in the vicinity of the Earth. And because hydrogen emits ultraviolet light at a wavelength called Lyman alpha, he looked to measures of Lyman alpha taken by interplanetary probes for a sign of the presence of small comets.

Donahue reasoned that any brightness in the Lyman alpha greater than that normally produced by hydrogen atoms coming into the solar system from interstellar space could be attributed to the small comets and would be easily seen. But, he pointed out, spacecraft such as Pioneer 10 and Voyagers 1 and 2 failed to find even a small increase above the sky background rate as they left the vicinity of the Earth on their way to rendezvous with the outer planets. Donahue proposed that the only way to stifle the large production of hydrogen from the small comets would be to cover them with thick dust mantles. But the problem with thick dust mantles, as we already know, is that they would bring, said Donahue, a "forbiddingly large" influx of dust into the Earth's atmosphere.

My principal criticism, as a referee on the paper, was that the Lyman alpha glow due to the presence of the small comets would *not* be easily detectable. Lyman alpha is very difficult to measure. It is not at a wavelength that we can see with our eyes and it is hard to make the instruments that are sensitive to it very accurate. Whenever we measure the Lyman alpha it is against an ever-present background glow. This light results when the Sun illuminates the interstellar hydrogen that surrounds us. It is very bright relative to the dim glows we expect to see from the small comets. Add that to the fact that the Sun's illumination of this hydrogen constantly changes and you have a very difficult measurement.

Lyman alpha is a tease and I had long ago learned to stay away from her. The work of Ralph Bohlin, one of my former undergraduate honors students, had warned me. Bohlin was a bright young guy. He had gone on to obtain his Ph.D. from Princeton and then joined the Laboratory of Atmospheric and Space Physics at the University of Colorado. Bohlin wrote a paper pointing out some of the discrepancies in the measurement of Lyman alpha in a comparison of data from three spacecraft, Mariner 6, Mariner 9, and OGO-5. When I looked at his paper and those of others on the subject and saw those discrepancies, I realized I would be in trouble if I based my small comet figures on Lyman alpha.

Bohlin's work had warned me. If I was going to propose that water was coming off these comets into the interstellar medium, then I ought to stay a long way away from Lyman alpha measurements. I would have to get the information some other way. So I did an end run around Lyman alpha. I went to my own specialty for an answer to the amount of hydrogen in interplanetary space. I realized that if the hydrogen from the small comets was out there, it would not remain a neutral atom very long. Either a solar wind proton or the ultraviolet light from the Sun would strip off its electron. Once it was ionized this hydrogen atom would get picked up in the wind of protons and electrons from the Sun and eventually be swept out beyond the planets. So I looked to the concentration of stripped-down hydrogen in the solar wind to determine the upper limit for how much hydrogen the small comets could produce in interplanetary space. I used this upper limit to estimate the outgassing of water from the small comets and found there should be some excess hydrogen light. But at most it should be only two to four times greater than the upper limit given by Donahue from the Voyager spacecraft.

Then in the spring of 1987 Donahue went back to the Voyager data to assure himself that no excess hydrogen light had been detected by Voyager. He hoped to deal the small comets their final blow. But instead he found this great cloud of hydrogen around the Sun in the vicinity of the Earth. It was this discovery that had led him to propose a new population of comet-like objects in interplanetary space. The proposal shocked everyone, I think, but no one more than Donahue himself.

When I finally ran into Donahue at the conference in Vancouver he let me see the paper that had caused the uproar. He called his objects cometesimals. These are distinctly different from small comets. Donahue's cometesimals are the rocky cores of comets in the making. They are baby comets. My small comets are comets in their own right. They are snowy on the inside and surrounded by a protective crust like Halley's, only much smaller.

I had been very careful in naming these objects. I could have called them cometoids, but that sounded too much like hemorrhoids. Some people have called them mini-comets or CLOs, meaning Comet-Like Objects. But others, aware that comets normally take their names from their discoverers, insisted on calling them FSCs, after the three names on the original papers, Frank, Sigwarth and Craven. Still others have jokingly referred to them as FSCs because they are Frank's Small Comets. I think the term small comets is best.

Donahue's cometesimals were remarkable creatures. They were ingeniously designed so that, unlike my own small comets, it was their first trip through the solar system. The cometesimals were new bodies with a nice, pure-ice coating. Just enough, in fact, that by the time they reached the Earth's orbit the Sun would have melted all the ice away. I guess it was possible to do that. With a little imagination one can construct almost anything. But there must have been good quality control somewhere out there while his cometesimals were being painted with a uniform coating of ice, the way peanut M&Ms are evenly coated with chocolate and sugar. Donahue and his colleague at the University of Michigan, Tamas Gombosi, and Bill Sandel of the University of Arizona, wrote a paper on the subject for *Nature*.

The ultraviolet spectrometer aboard Voyager 2 had taken a series of seventeen observations of interstellar Lyman alpha from hydrogen as the spacecraft sped away from Earth in 1977. The figures showed a rapid decline in Lyman alpha as the spacecraft travelled from a point about 7 million miles away from the Earth to one that was about 140 million miles away. The initial measurements showed an excess brightness that ran about one-third greater than the light due to interstellar hydrogen but eventually dwindled to nothing as the spacecraft receded from the Earth. Clearly this indicated the presence of a cloud of hydrogen near the Earth's orbit. The most reasonable explanation for it, Donahue concluded, is that a large number of comets were losing their ice as they passed through the inner solar system.

Donahue's cometesimals bore a striking resemblance to my own small comets. His had a diameter of 26 to 164 feet. Mine measured 40 feet on average. His weighed 30 to 30,000 tons. Mine weighed about 100 tons. His comets were small enough to have escaped detection by ordinary telescopes. So were mine. But his were boulders with thin ice jackets. Mine were snowballs with thin carbon mantles.

Where we differed the most, however, was in the number of objects. His proposal called for a hundred million fewer comets than mine. His rate, Donahue explained, was consistent with the number of craters on Mercury, Mars and the Moon. Donahue had, in fact, limited the number of his cometesimals through a count of the number of lunar craters that such objects were likely to produce. He found that his cometesimals would account perfectly for every lunar crater measuring up to about 5,000 feet in diameter. These craters were previously assumed to have been formed by stone or iron meteoroids.

Donahue's cometesimals would have less of an impact on Earth than my small comets. His baby comets would hit our planet only once in every eight years, while the largest ones would impact about once in every 10,000 years. My small comets are hitting the Earth's atmosphere at a rate of about twenty per minute and have deposited enough water here over time to have filled the ocean basins. But the total cosmic rainfall from Donahue's cometesimals would be insignificant. They would not have brought the ocean waters to the Earth.

His discovery had not caused Donahue to change his mind about my small comets. His was the simpler hypothesis and therefore the best, he concluded. In a far more detailed exposition of his proposal that he prepared for *Icarus*, Donahue actually invoked the rule of Ockham's razor in support of his cometesimals. William of Ockham was a 14th century scholar who said that "entities must not needlessly be multiplied."

Modern science has since interpreted these words to mean that if two theories fit equally well all the observed facts, the theory requiring the fewest or simplest assumptions was more likely to be valid. Most scientists do not use this razor because its blade often cuts the wrong way.

Donahue's call for Ockham's famous razor to decide between our two proposals puzzled me. But it infuriated George Wetherill, the world-famous scientist from the Carnegie Institution of Washington who also refereed Donahue's paper for *Icarus*. "Yuk!" wrote Wetherill. "This is an argument that is useful only to an uninformed audience whose poor educations left them with the impression that a medieval scholastic understood inversion theory better than modern scientists. Usually it is used as an implicit admission of a lack of a good reason for rejection of someone else's hypothesis, but to have to use it against Frank's holes???!" Obviously, Wetherill was not fond of my small comets either.

It seemed to me that Donahue had completely reversed his previous position. He might be denying the existence of my small comets, but at the same time he had entered the cometary competition with an icy celestial population of his own. I was no longer alone. I was being faulted for numbers that were too high. Donahue was being faulted for numbers that were too low. Some people were beginning to think that the truth might lie somewhere in between. "I think that Donahue's work has had the effect of giving great credence to the existence of this class of objects," Dessler said at the time. "Now they are only dickering over the numbers." Never mind that our numbers were a hundred million apart.

Despite my excitement, I was wary of the tease of Lyman alpha and I wondered why the measure of hydrogen light from Voyager failed to agree with the measurements from other spacecraft. When Halley's Comet came rolling through the solar system, various satellites were turned towards it. One that could look at the hydrogen gas around the comet was Pioneer Venus. Ian Stewart of the University of Colorado headed that effort. We also looked at Halley's with Dynamics Explorer. My colleague John Craven did the analysis. What we learned from these observations was that the hydrogen envelope around Halley's is enormous.

The measurements from Dynamics Explorer and Pioneer Venus agreed. This meant that the two instruments reported back to Earth the same brightness for the same object. This was also true for the background light from the interstellar hydrogen that was falling into the solar system. But the Voyager instrument, the one that Donahue had used to search for cometary water, reported glows from the same interstellar hydrogen that were two or three times brighter than these. That is a big difference, especially if you are going to claim you can measure something to within a few percent and you fail to agree by 200 or 300 percent with what everyone else says is going on. Donahue was aware of the calibration problem. "But," he said, "who's right?"

Dessler got angry when Donahue's paper appeared in *Nature*. Donahue had not even submitted it to *Geophysical Research Letters*, where the debate had been taking place. The *Nature* paper got a lot of publicity, and this no doubt made Dessler very jealous. But, in any case, it did not take long for everyone to repeat Donahue's blanket statement that this newly found hydrogen had nothing to do with my small comets. But if there was all this hydrogen gas out there, how could anyone really know what it came from? It could come from automobiles or sick cows. The real significance of Donahue's finding was that there were objects out there that had never been detected before. It did not matter whether they were small comets, cometesimals, automobiles or sick cows.

Chapter 16

The Verdict

I felt obliged to respond to Donahue's discovery of cometesimals. His paper had appeared in the British journal *Nature*, one of the great keepers of current wisdom in science in the Western world. Current wisdom is what a large body of scientists in a given discipline believes and holds sacred. Current wisdom is usually—but not always—correct. Most scientists will jump through hoops to get their papers published in *Nature* and its American counterpart *Science*, as both journals are very popular and widely distributed. But I have had little to do with either one.

I was published in *Nature* as a junior co-author with James Van Allen some three decades ago. And recently I had a paper on a subject not related to small comets rejected by *Science*. It was returned to me without review or comment. For the keepers of current wisdom it seems that the small comets are now the equivalent of bad breath. It matters little to me. I prefer to publish in journals where you can generally count on the competence of reviewers; this has not been my experience when the journal is a keeper of current wisdom. For my work, a competent reviewer means a peer with good skills in basic science.

I am not very good at responding effectively to folklore and faulty current wisdom. I was reminded of that when I tried to respond to Donahue's claim that the excess hydrogen found with the Voyager observations had nothing to do with the small comets that I proposed, even though they are what led him to search for the hydrogen in the first place. What troubled me was his claim that only his model, his number of objects, could account for the hydrogen. The simple fact is that a large number of object sizes and types could account for the Lyman alpha, the dim glow from the thin hydrogen gas in interplanetary space.

So I decided to write a paper saying, in essence, that the small comet hypothesis could also satisfactorily explain Voyager's Lyman alpha emissions. In fact, almost everyone in the United States could have their own model of the number of objects and how their surfaces behaved that would give the same hydrogen light as seen by Voyager. There was nothing miraculous about the contents of my paper. Several colleagues recommended that I submit it to *Nature*, given that Donahue's paper had been published there. My paper was an alternate interpretation of the excess Lyman alpha brightness and *Nature* seemed to be the appropriate place for it. I should have known better.

The referees assigned to my paper made it quite clear that they were not interested in seeing anything else on the subject. The first referee thought that *Nature* should decide "whether the small comet proposal has been so discredited that further discussion of it" was even appropriate. More specifically, he wanted me to explain how the comet mantles would adequately suppress water production and what happens to the water from the comets inside Earth's orbit. I had tackled the first problem earlier in *Geophysical Research Letters*. The second problem I took care of in my revision. They had me extend the computations far beyond what Donahue had done in his paper and I satisfied all the requirements.

The first referee also raised one of the most ridiculous objections I had seen in some time. He criticized our paper because we had not proven that ice melts when it is heated above its melting point. What world had I entered? How do you answer such a question? If you heat ice above its melting point will it melt? So I answered that in my experience every time I had heated ice above its melting point it had turned to liquid water. But perhaps the referee had another experience. It was quite apparent that his emotions far exceeded his common sense.

The second referee agreed with the first, except on the melting point of ice. He wrote that I had failed to provide "a compelling proof of the mini-comet hypothesis." That, of course, was not the point. He noted that I had referred to many of my arguments from previous replies to criticisms of our work and that these replies had not been reviewed by others before publication. This led him to conclude that we had "built a self-referenced, unreviewed, house of cards that will fall to pieces if even one of their previous replies is incorrect." I had always said myself that only one bit of evidence was needed to prove the small comet hypothesis incorrect. That in fact was what the debate was all about. But so far no one had been able to come up with it.

The comments by the second referee were also personally insulting. This is often the case with less than top-notch referees when they fail to understand the contents of a paper or are emotionally attached to current wisdom. The fact that they are anonymous sometimes brings out the worst in them, they seem to relish this sort of lion-hunting in a cage. The referee said I was "using hand-waving arguments and approximations," and on seeing my reply to the first referee, accused me of Nixon-like behavior. I was stonewalling, he said.

Thinking the British editors of *Nature* so naive as to be unfamiliar with the term, the referee provided an in-depth explanation: "Stonewalling is the process of repeatedly denying an unacceptable statement in the hope that the critic will get tired and give up. It was used most notably by the Nixon administration when they denied any involvement in Watergate, no matter what evidence was presented to the contrary." This was clever, but it was not science.

So the paper went back and forth, back and forth, and then came the final decision from the editor in charge of the *Nature* office in the United States, David Lindley. He rejected the paper. "At this stage," he wrote, "we are satisfied that we have taken a representative poll of current opinion in this field, and the verdict goes against you." I could not believe it. My paper was rejected not on the basis of something being wrong with the arithmetic or something being wrong with the science, but because *the vote went against me.*

I was astonished by the arrogance of Lindley's statement. The verdict had gone against me. You would think that they could have voted the small comets into office had they been popular enough. But on second thought I must say that I became very enthusiastic about this "physics by vote." This kind of voting process would allow us to dispose of the laws of thermodynamics. Perhaps next year the community would revoke them and I could look forward to becoming younger. It would also mean a quick end to the energy crisis.

It was mind boggling. I thought that science was what was correct and verifiable by experiment and observation. Now, suddenly, it was being done by the vote of the members of a given community. This was a very distressing learning experience. To think that Newton's laws might be *voted* out. Perhaps someone would take Ockham's razor and slash them from our universe.

I had one other brush with *Nature* during the small comets controversy. It occurred just as my own paper was being reviewed at *Nature*. They decided to prepare a story on the small comets for the "News and Views" section of the journal and Paul Feldman was asked to write it. Feldman is a professor of physics and astronomy at Johns Hopkins University in Baltimore and is one of the country's leading experts on comets.

At the time, Feldman and I were competing for an imager on the NASA Polar spacecraft due to be launched in 1993. The camera had nothing to do with small comets. The competition was intense because the opportunity to fly an instrument into space these days is so rare. Feldman is a good physicist. He is bright and energetic. He was not too familiar with the small comets, but he understood what they meant for the solar system and for science. Feldman accepted the assignment.

Feldman told me that he would treat the small comets in a kindly fashion but the story he wrote was downright nasty. Feldman showed a masterful use of the thesaurus, but he put absolutely no science into it. He apparently relished poking fun at me. He compared me to René-Prosper Blondot, who in 1903, was the only person able to observe the non-existent N-rays.

He also labeled as "Velikovskian" the way small comets were able to account for so many of the solar system's mysteries. He was referring to the late Immanuel Velikovsky and his controversial theories of recent catastrophes in the solar system. Velikovsky's use of the world's myths and legends as proof of his theories has long been ridiculed by science. Feldman also accused me of constantly "reengineering" or adjusting the properties and characteristics of the small comets as I responded to each of my critics. But Feldman could not have read my papers because there was, in fact, no change.

I did not mind. Feldman is a friend. It is difficult for me to be angry with him since he is as child-like as I am when it comes to things of this sort. I got what I expected. We compete nose to nose. He wins some. He loses some. That's life. He got to write that article for *Nature*. I got to do the imager on the NASA Polar spacecraft. Feldman told me privately afterwards that he would never write anything like that again.

I am a member of the American Association for the Advancement of Science and have a profound respect for its journal, *Science*. It claims to represent "the advancement of science." So the news reporters for *Science*, such as Richard Kerr, wield an awesome amount of power. They are given a free hand in their work.

Kerr first approached me to do a story on the small comets at the American Geophysical Union meeting in San Francisco in December of 1987. I told him that I was not particularly interested in having the topic covered in *Science* because the debate was not going on there. Kerr is a very knowledgeable fellow. He has a tremendous task, of course. He has to come up with a couple of articles a week on a myriad of topics.

But write about the small comets, he did. He wrote three stories in less than a year. Kerr thought the idea of small comets absurd. It was, of course, his right to think so. If I were in his shoes I would probably say the same thing. The odds were not in my favor. Why gamble the other way? It did not bother me that he so openly expressed his opinion on the subject. What irritated me was that his articles contained factual inaccuracies. It does not matter when this happens in a little newspaper. But when you get a periodical as prestigious as *Science* publishing something that countains inaccuracies, something has gone seriously wrong.

I let it go the first time. But his second article had even more errors than the first. So for the first time in my life I wrote a letter-to-the-editor trying to set the record straight. It was a futile effort but it did allow me to vent some frustration. There were several factual errors in the Kerr stories. The first error was that he made John Craven a graduate student. John Craven is a senior research scientist. He has not been a graduate student for about twenty-five years. He also stated that the small comets with their protective crusts would release ten million times more hydrogen in interplanetary space than observed, when in fact this was the value for comets with no protective crusts—a big difference.

This information—misinformation rather—was apparently acquired through hearsay, which is not unusual for such a heated subject. Kerr's second article also mentioned that a workshop had been called in Boulder because of Donahue's discovery of the cometesimals. Kerr was not there. I was. That was not the purpose of the meeting. Nor did he mention the very significant outcome of the meeting: that the most important goal for the Galileo spacecraft during the initial part of its journey should be to try to make measurements of the Lyman alpha light from hydrogen in the Earth's vicinity. It was just sloppy reporting. This type of journalism is not a service to readers. I was very disappointed. But I learned that when it comes to heated topics, there was obviously no guarantee of accuracy and fairness, even in the scientific press.

The letter had some effect. Kerr began calling around. He learned that Craven was indeed not a graduate student. *Science* printed a correction in the tiniest possible print in one of its subsequent issues. No other mistake was corrected. But Kerr's next article on the subject showed a marked improvement. I have to compliment him because in the end he tried to correct things. He tried to find out more about the issue. But the whole episode made me question the other news in *Science*. Could I still trust them to tell me what was going on, especially on a dispute I knew little about? Would I get the real picture when I read about a controversy in medicine? It was another loss of innocence on my part.

But my biggest disappointment with the press came not with these big fish but with one of the smaller fish in the scientific pond. I belong to the American Geophysical Union and have been to its meetings since I was very young. I have actually been chosen from its large group of members as a Fellow of the Union.

The Union has a weekly paper called *EOS*. It reports everything from Nobel-Prize winner Hannes Alfvén getting the Union's top honor down to who stubbed their toe last Tuesday. The small comets issue has involved more people and caused more controversy and more discussion than most issues that come up in the Union. But one thing that hurt me—and most of these things do not bother me—is that I was so heavily censored on the small comets issue that not one word has ever been written in that newspaper about the topic. Not a single word. Others have asked about it and they are as amazed by the situation as I am. Meanwhile, reports on my other work have continued to appear in *EOS*. Perhaps the editors thought that the issue of small comets was of no importance to the studies of the Earth.

Sour grapes? Yes, in part. It has been a very frustrating experience for me. People tend to forget that scientists are human, too. Though I cherish the world of science and its complexity, my childhood on the plains of Kansas and in a small rivertown on the Mississippi was spent with a different kind of people, in a different kind of world. In the course of this debate I would sometimes return to this other world to forget my frustrations with the intrigues and the public ridicule surrounding the small comets. I would jump in my truck or my car and head the sixty or so miles east to the Mississippi and its rundown taverns and smoke-filled pool halls.

The real attraction for me there was the people. They are truck drivers, mechanics, businessmen, even Hell's Angels. These people accept you as you are with them and not for what you might be elsewhere. No backdoor scheming, no abrasive confrontations, just a simple and direct interaction. Two very different worlds. There is no basic difference in intelligence, only a matter of education, which is lost here anyway. A short while spent with these people renewed my contact with the basic realities. It was a deep pleasure, a moment away from the paddling line.

Chapter 17

The Competition Fizzles

The sides in the comet competition were clearly drawn. It was Donahue versus Frank. The inner solar system was not big enough for the both of us. We knew that a sign in the heavens could declare one of us the winner, but this sign unfortunately was not visible from the Earth. What we needed could only be had from interplanetary space: a portrait of the hydrogen light from these objects in the Earth's vicinity.

If the cometesimals were real, this portrait would resemble a shell of hydrogen left over from the vaporization of these objects by the Sun as they reached the Earth's vicinity. Consequently, this shell would be centered on the Sun and located in Earth's orbit. If the small comets were real, the portrait would instead show up as a torus, or ring, because the small comets, unlike Donahue's cometesimals, are confined near the planes in which the planets travel about the Sun.

Donahue and I knew that this observation of the Lyman alpha light in interplanetary space could only be made with the proper spacecraft. That spacecraft, we realized early on, would be Galileo. It was scheduled to be launched by the space shuttle Atlantis in the fall of 1989. Galileo was a mission to Jupiter. It was designed to drop a probe into the planet's atmosphere while the orbiter itself returned pictures and other data on the planet's atmosphere, the powerful radiation zones that encircle it, as well as its many moons. Because Galileo missed its scheduled flight date due to the Challenger disaster, it would have to swing by Venus and Earth for gravitational boosts to reach Jupiter. It was like an interplanetary pinball game. But instead of two years, it would now take six to get to Jupiter. The spacecraft was never meant to go to Venus, but now that it had to make that flight, no one wanted the side trip to be a waste.

There was much work to do. In June of 1988 some of the scientists on the Galileo mission gathered for a workshop at the University of Colorado in Boulder. I was there. Donahue was there. Altogether more than fifty scientists had been invited. Practically everyone interested in interplanetary hydrogen attended. A few were curiosity seekers looking for a show of tempers. Don Hunten of the University of Arizona chaired the meeting. Its sole purpose was to discuss and establish priorities for Galileo during the Venus to Earth portion of the mission. There were a dozen different experiments on the table but only a limited amount of data storage available on the Galileo tape recorder. During this time we would argue whether the spacecraft could do anything useful on the long-debated topic of just how much hydrogen there was in interplanetary space. The fate of the small comets and the cometesimals hung in the balance.

The best time to look at the hydrogen light is on the way to Venus and back again and Galileo was well equipped to make this observation. There would be two ultraviolet instruments on board and each was capable of measuring the ultraviolet light from hydrogen known as Lyman alpha. One of them, the University of Arizona instrument, would be making observations at a very low rate, but these would be transmitted to Earth immediately. Observations from the other ultraviolet instrument, the pride and joy of the University of Colorado, however, would have to be stored on the spacecraft tape recorder. These observations could not be transmitted to Earth until the spacecraft had returned from Venus, about one year after the launch of Galileo.

The Universities of Colorado and Arizona are fierce competitors in this ultraviolet business, but curiously enough Colorado had helped Arizona get its instrument on board. Arizona's ultraviolet imager was actually an unused spare from the earlier Voyager mission and was added to Galileo when the Challenger disaster delayed the mission. I, too, had argued for it, and John Casani, the Galileo project manager, had also been extremely supportive. It did not matter to me that this was one of the groups that had hurled the most severe criticism at me during the small comet debate. I thought that it was in the best interest of science. I felt it would be very worthwhile for imaging the great doughnut of gases and plasma that encircle Jupiter, so I worked hard to get their instrument on board.

It was an easy decision. Science first. This is the tie that binds. And its threads run throughout the Galileo project. Casani, who I consider to be the world's best space project manager, is interested in advancing science as well as implementing a successful spacecraft. We had first worked together when I was twenty years old. He was a younger engineer on the first lunar probes. He would bring the tiny Pioneer 3 and 4 spacecraft to Iowa and we would calibrate them with an x-ray machine. James Van Allen was the principal scientist. It would usually take all night and at the end we would sit on the steps outside the lab and, with a beer in our hands, watch the Sun rise.

Just a few days before the Boulder workshop, two scientists from the University of Arizona dropped a bombshell. Donald Shemansky of the Lunar and Planetary Laboratory had been amassing spacecraft observations and began to have doubts regarding Donahue's claims of an excess of Lyman alpha in interplanetary space. So he had asked his student Doyle Hall to compare Donahue's results with the results from their own independent inspection of these measurements. Doyle did not find an excess in Lyman alpha light and his upper limit was only a small fraction of that claimed by Donahue, if indeed there was any extra light at all. Hall and Shemansky concluded that Donahue had erred.

I read a copy of Hall and Shemansky's as-yet-unpublished paper that was circulated at the meeting and I found their claim of how well they could measure the Lyman alpha light a little overconfident. But I think they had carefully and correctly determined that Donahue had indeed made an error. Donahue said that he would go home after the meeting and redo his calculations. But everybody there seemed convinced that he had made a mistake. The news did not bode well.

But in the end, Donahue's error did not affect the group's decision. The group recommended, in a paper known as the "Hunten Report," that the most important thing for Galileo to do between Earth and Venus was to search for the hydrogen from unknown objects. "The primary objective has high priority whatever the actual comet population," wrote Don Hunten. Everyone realized that this was the perfect opportunity to get good measurements of the Lyman alpha light in interplanetary space. "Controversial or not," Hunten wrote, "the [small comet] hypothesis is certainly interesting and deserves to be tested."

After the Boulder meeting Donahue checked his calculations and found that Hall and Shemansky were correct. "I blew it," Donahue told *Science*. He explained that a graduate student had committed the error. The student had made a slip in a decimal point. It had been a clerical error, in other words. But I looked at the author list on the original Donahue paper. There was Tamas Gombosi, a senior research physicist at Michigan, and Bill Sandel at the University of Arizona. But there was no graduate student. In the acknowledgments Donahue thanked Tom Ahrens, but he is a professor at Caltech, and Harold Masursky, who is well known for his work on the channels of Mars. I was surprised that a graduate student who had so much to do with the result had not even been acknowledged.

But perhaps it *was* a graduate student's fault. Donahue, at the time, was very busy as chairman of the Space Science Board. This is a tremendous responsibility, so perhaps he had to rely on others and did not have a chance to directly oversee the data analysis and computations. But in matters like this, the advisor is not only responsible for checking the work, but also, ultimately, for the conclusions. It was a very embarrassing situation.

The final version of the paper that Hall and Shemansky submitted to *Nature* in July began with a statement saying that the excess hydrogen light found by Donahue with Voyager was about ten million times less than what would be expected from my small comets. I could not believe it. Donahue had made the error but right away I was being put on the spot. The Lyman alpha light would have been ten million times brighter only if the small comets had no mantle whatsoever. Never did I say that

Geophysical
Research
Letters

**APRIL
1986**

**volume 13
number 4**

AMERICAN GEOPHYSICAL UNION

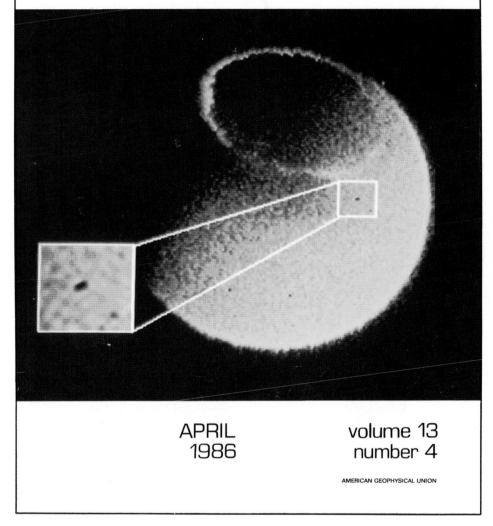

The cover of the scientific journal which announced the discovery of the small comets. The ring at the top of the ultraviolet image is the northern auroral oval. The inset is an expanded view of an "atmospheric hole" produced by a small comet. This black spot actually represents a cloud of water vapor about 30 miles in diameter at an altitude in the range of 200 to 600 miles above the Earth's surface. This is what remains of a small comet when the house-sized chunk of water-snow is vaporized above the Earth's atmosphere.

The data that led us to propose the existence of the small comets and their present-day bombardment of the Earth came from an ultraviolet imager on this NASA satellite, Dynamics Explorer 1. It was launched on the 3rd of August 1981 from Vandenberg Air Force Base in California. The satellite continues to orbit the Earth and continues to transmit evidence of the small comets.

The images from the Dynamics Explorer satellite are not taken "at once" like a camera, but produced one line at a time. This means that when the satellite is at a high altitude, the black spots that appear in adjacent lines (see inset) will represent the motion of a single object as it plunges towards, and finally into, the Earth's atmosphere. A study of these "apparent motions" suggests that the objects—these water vapor clouds produced by the vaporization of the small comets—are travelling at about 10 miles per second just above the Earth's atmosphere.

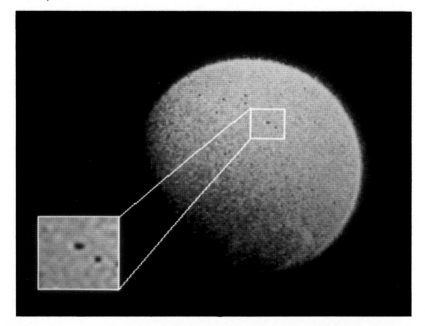

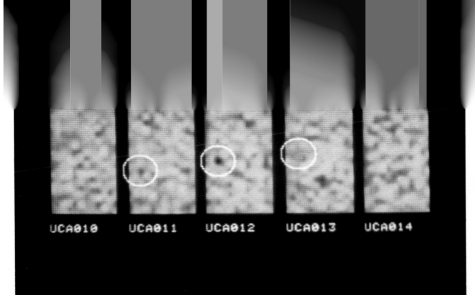

UCA010   UCA011   UCA012   UCA013   UCA014

The gradual appearance and disappearance of a single atmospheric hole (circled), which is caused by a small comet, in a sequence of images taken 72 seconds apart. These images reflect the transient nature of the water vapor clouds, or remains of the small comets, as they plunge into the Earth's atmosphere. A few other atmospheric holes can be seen in the images.

The cometary water clouds should have a bright leading edge from plowing through the thin gases in the upper regions of the atmosphere. The global image (left) shows the bright front-half of these objects as they collide with the Earth's tenuous atmosphere at high altitudes. The inset in the center is enlarged six times the original size. The left half of the inset shows a pattern of three stars. The arrow in the inset indicates the bright leading edge of a cometary water cloud.

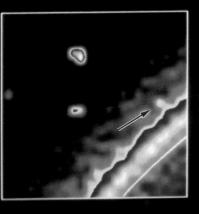

WCI334   81 /285 13:32 UT
FILTER 123W   3,58 Re
GG: 66 N, 117 E, 21:35 LT
GM: 60 N, 188 E, 20:59 LT
120 SCAN LINES (29 -148)

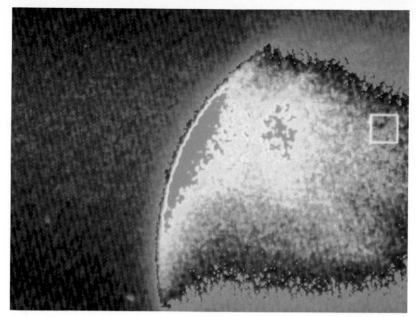

An ultraviolet imager on another Earth-orbiting satellite, the Swedish Viking satellite launched in February of 1986, has also shown evidence of atmospheric holes caused by the small comets.

In November of 1987 and January of 1988 Jet Propulsion Laboratory physicist Clayne Yeates searched for the small comets using the Spacewatch Telescope at the Stewart Observatory at Kitt Peak, Arizona. By using the predicted motion of the small comets from observations of atmospheric holes, Yeates managed to get the first optical pictures of the small comets with this specially equipped telescope. The large streaks in the image are star trails, but a faint, small comet trail appears as well (the slanted line in the outlined box). These small, dark, fast-moving objects have so far only appeared in images taken at the very limits of detection.

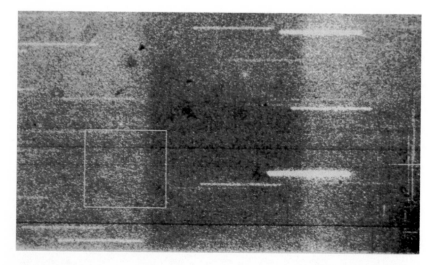

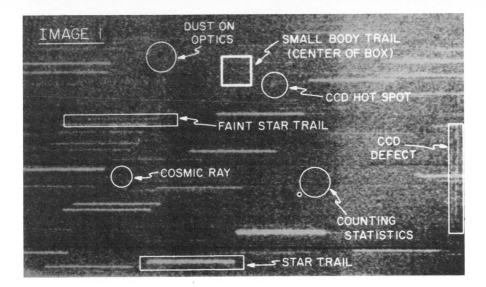

The telescope images of the small comets contain a lot of "noise," some of which is identified in this frame. A small comet appears as a faint trail in the box at the top of the image. This is the first in a series of consecutive exposures taken by University of Arizona astronomer Tom Gehrels in April of 1988, and subsequently analyzed for the presence of small comets by Clayne Yeates, John Sigwarth, and myself.

The same comet seen in the previous exposure is seen in the box in this exposure, which began 36 seconds later. Taking consecutive frames of the same object is the standard of proof in astronomy.

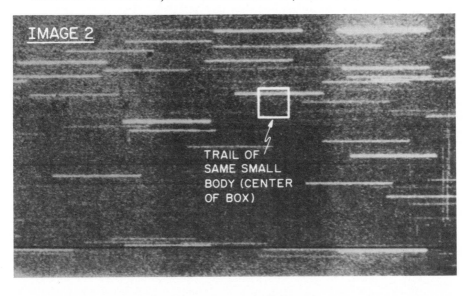

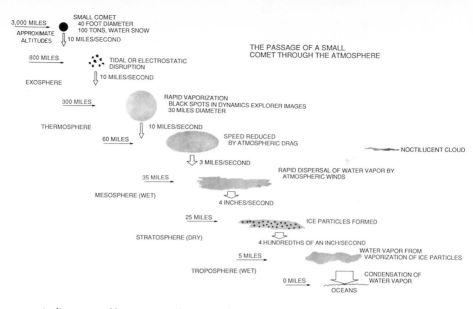

THE PASSAGE OF A SMALL
COMET THROUGH THE ATMOSPHERE

3,000 MILES — APPROXIMATE ALTITUDES

SMALL COMET
40 FOOT DIAMETER
100 TONS, WATER SNOW
10 MILES/SECOND

800 MILES

TIDAL OR ELECTROSTATIC DISRUPTION

10 MILES/SECOND

EXOSPHERE

300 MILES

RAPID VAPORIZATION
BLACK SPOTS IN DYNAMICS EXPLORER IMAGES
30 MILES DIAMETER

THERMOSPHERE

10 MILES/SECOND

60 MILES

SPEED REDUCED
BY ATMOSPHERIC DRAG

NOCTILUCENT CLOUD

3 MILES/SECOND

35 MILES

RAPID DISPERSAL OF WATER VAPOR BY
ATMOSPHERIC WINDS

MESOSPHERE (WET)

4 INCHES/SECOND

25 MILES

ICE PARTICLES FORMED

STRATOSPHERE (DRY)

4 HUNDREDTHS OF AN INCH/SECOND

5 MILES

WATER VAPOR FROM
VAPORIZATION OF ICE PARTICLES

TROPOSPHERE (WET)

0 MILES

CONDENSATION OF
WATER VAPOR

OCEANS

A diagram of how a small comet deposits its water on the Earth. These objects have supplied the Earth with enough water over the past four billion years to fill the oceans.

These high-altitude noctilucent clouds are something of a mystery to atmospheric scientists. Not enough water is thought to evaporate from the Earth's surface and reach altitudes of 55 miles to form these clouds. But if the small comets are bringing the water in and depositing a little of it in the upper atmosphere, then the mystery has a very simple solution.

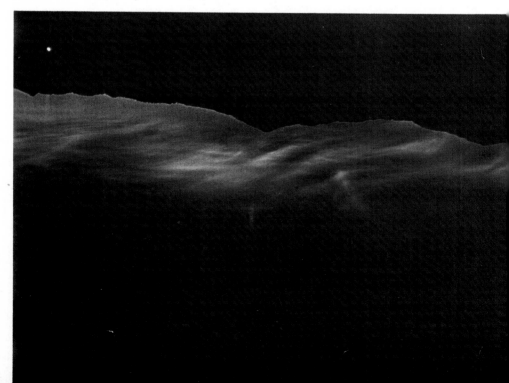

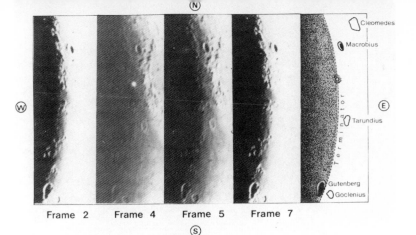

Cleomedes

Macrobius

Terminator

Tarundius

Gutenberg
Goclenius

Frame 2   Frame 4   Frame 5   Frame 7

If the small comets are bombarding the Earth's atmosphere they are likely to be falling into the Moon as well. For centuries, strange glows and lights have been reported on the Moon. In 1985 Greek astronomers captured such a transient flash in frame 4 of a sequence of exposures. I believe that the impact of a small comet on the Moon would produce just such a momentary glow.

The outer planets are also likely to be bombarded by small comets. The surface of Mars is rippled with channels and tributaries suggesting that water once flowed there in great quantities. This water may well have come from an influx of small comets.

The large, known comets, as well as the small comets, are thought to come from a large region that lies beyond the orbits of Neptune and Pluto. Evidence that such a belt of matter might lie outside the known planets comes from many sources, one of which is the telescopic observations of other stellar systems such as this one, known as Beta Pictoris. Though planets are thought to be forming at the very center of this disk, a good deal of matter lies far beyond it. Likewise, a large amount of the matter in our solar system probably lies beyond the known planets.

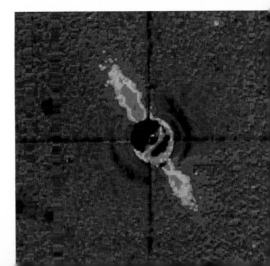

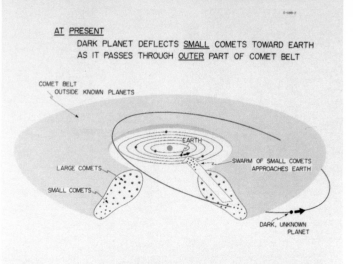

AT PRESENT

DARK PLANET DEFLECTS <u>SMALL</u> COMETS TOWARD EARTH
AS IT PASSES THROUGH <u>OUTER</u> PART OF COMET BELT

To explain the constant bombardment of the Earth by small comets, I have proposed the existence of a large, dark, as-yet-undiscovered planet that regularly passes through the outer part of the comet belt where the small comets are thought to be located. These transits of the belt by the Dark Planet send swarms of small comets streaming into the inner solar system and the Earth itself.

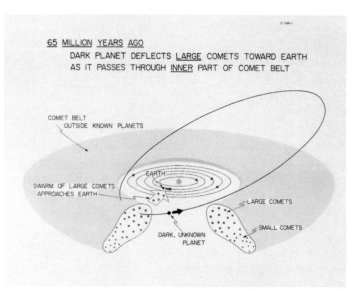

65 <u>MILLION YEARS AGO</u>

DARK PLANET DEFLECTS <u>LARGE</u> COMETS TOWARD EARTH
AS IT PASSES THROUGH <u>INNER</u> PART OF COMET BELT

I believe that the eccentric orbit of this Dark Planet causes it to transit the inner portions of the comet belt—where the large, known comets are thought to come from—once every 26 million years or so. When it happened 65 million years ago, the Earth was bombarded by a swarm of large comets that created a "nuclear winter" scenario on Earth, killing off many living species, including the dinosaurs. The next swarm of large comets, caused by such a transit, is not expected to reach the Earth for another 12 million years or so.

the comets would have no mantles. But I suppose Hall and Shemansky had to come up with something startling—even if it was incorrect—in order to get published in *Nature*.

In their eagerness, Hall and Shemansky went on to cite a non-existent paper, which I had supposedly written for *Geophysical Research Letters*, on how Donahue's Voyager data fit my small comets proposal. This was the paper I had submitted to *Nature*, not *Geophysical Research Letters*, and it had been rejected. I had written to Shemansky in August informing him of this incorrect citation and asked that it be corrected. But it never happened. The whole shooting match—incorrect citation and misleading figures on Lyman alpha from the small comets—was published in *Nature* in September.

The mistake rapidly snowballed. Richard Kerr, writing for *Science*, picked up the information for his story from the Hall and Shemansky paper without bothering to check with me to see if it was right. This imaginary "off by a factor of ten million" would later come back to haunt me at the next Galileo Project Science Group meeting. It was quite a mess.

I would think the Hall and Shemansky affair caused Donahue the greatest disappointment. His cometesimals had nothing left to stand on. But I have to give him credit. At least he had gone in there and tried to resolve the issue. Few had done anything at all. But he had worked very hard at it. So it did not surprise me that, in the end, Donahue tried to take my small comets down with his cometesimals. When Hall and Shemansky proved him wrong, Donahue said, in effect, "Well, that kills Frank's small comets, too." Never mind that when he had first proposed his cometesimals, he had insisted that they had nothing to do with the small comets. It was a clever move by Donahue. People tell me that was not fair. But fairness had nothing to do with it. There was nothing dishonest about it either. People fail to understand that science, quite simply, is a savage little game that we play with a great deal of devotion.

The final go-ahead to search for the hydrogen from a sea of small comets had to be made by the entire group of Galileo scientists at a Project Science Group meeting at the Jet Propulsion Laboratory in Pasadena the following January. The membership of the PSG, as this group is known, is truly impressive. It includes James Pollack of the NASA Ames Research Center, who helped develop the notion of "nuclear winter"; James Hansen, head of the NASA Goddard Institute for Space Studies, who believes that global warming is happening now; James Van Allen, who is the recent recipient of the Crafoord Prize, considered the equivalent of the Nobel Prize in space science; Andrew Ingersoll of the California Institute of Technology, who is best known for his insight into the weather of planets other than Earth; and Christopher Russell of UCLA, who has instigated the heated debate on whether or not there is lightning on Venus. Some of the members of the PSG have instruments on Galileo. Others simply intend to use the data from these instruments to develop their own ideas on the nature of Jupiter, its moons, and its immense radiation zones. I am a member of the PSG and have an instrument on Galileo that will look at the ions and electrons in Jupiter's radiation zones.

A PSG meeting lasts a day or two. The stresses and tensions increase as the meeting progresses. Strong personalities are involved and what one group wants to do with the Galileo spacecraft often conflicts with the desires of others. Galileo could not possible satisfy all of their demands, even though it is probably the most advanced space exploring robot to be flown before the turn of the millennium. But this is where the buck stops; all conflicts must be settled at these meetings. It is not a pretty process. The cheery smiles seen on the first day soon give way to faces etched with grim determination. As the tension builds, there are abrasive exchanges of insight and wit. But compromise and resolve must prevail in the end, as every bit of treasure is squeezed from the two-billion-dollar Galileo mission.

Everyone at this PSG meeting in Pasadena was fighting for resources. I was part of a long parade of scientists, all pitching their own projects for the Galileo mission. The competition for the limited data space on the spacecraft's tape recorder during the Earth-to-Venus-to-Earth leg of the journey was especially intense. Some people were arguing that the instruments should be used to study Venus lightning. Others wanted to use them to study the motions of the Venus atmosphere. Everyone was trying to get a slice of the pie. And, as expected, the idea of using the severely limited time on the tape recorder for observations of interplanetary hydrogen produced by the small comets quickly came under attack.

I got up to defend the taking of Lyman alpha measurements for interplanetary hydrogen. Mike Belton, who is in charge of the imaging on Galileo, pointed out immediately, for anyone not already aware of it, that I was the small comets person. Jim Pollack followed Belton. He asked why we were even discussing the small comets since no one had ever shown they were real. There is no supporting evidence for them, he said. Pollark was biased. We all were. If Galileo did not do the Lyman alpha measurements, Pollack could get his measurements on the Venus atmosphere.

Then Belton whipped out the Hall and Shemansky paper and said we did not need to go look for hydrogen in interplanetary space from small comets. He quoted Hall and Shemansky as saying that if the small comets were real then the Lyman alpha would be about ten million times larger than had been previously detected anyway. I expected that. We used everything we could against one another. But I pointed out that Hall and Shemansky's statement was incorrect. Belton got a little angry at me. I got up and asked him what else he was going to quote. There was no response.

I decided to sit down. There was no reason to stand up and face the wrath of this group in support of science to be done by instruments belonging to the Universities of Colorado and Arizona. It was a no-win proposition. I half expected some words on my behalf from my arch-critics at the University of Arizona. After all, they had gotten their instrument onboard Galileo only recently and, in part, due to my efforts. But to my surprise, it was Ian Stewart, in particular, and Charles Hord, both from the University of Colorado, who strongly supported the search for hydrogen in space. It was their direct support of a search for small comets that would assure time on the Galileo tape recorder. Ian Stewart, who was one of my most vocal critics, would lead the charge. Strange bedfellows, indeed!

Chapter 18

Why the Moon Doesn't Ring Like a Bell

Space should hold many clues to the existence of the small comets. Even before the issue of interplanetary hydrogen arose, the Moon had been the focus of the debate regarding the small comets. Everyone realized that the Moon presented the first real test outside of Earth for the existence of the small comets. The Moon is our companion here. It is the first outpost of that fortress known as Earth. Anything that happens to the Earth is likely to happen to the Moon, and vice versa. If small comets are bombarding the Earth, small comets must be pelting the Moon as well. Likewise if there is no evidence for small comets on the Moon, then small comets would have to go the way of green cheese and other lunar folklore.

Even before our original papers were published we had considered the Moon problem. The primary issue was this: If an infall of comets similar to what we proposed for the Earth hits the Moon, should we not have seen a far greater number of seismic impacts on the Moon than were observed with the detectors, called seismometers, placed on the lunar surface by the Apollo astronauts?

Between 1969 and 1972 seismometers were left on the Moon to measure ground movements due to moonquakes and meteoroid impacts. This network of instruments detected 12,500 moonquakes and 1,743 meteoroid impacts during its eight and a half years of operation. These instruments recorded, on average, a little more than one meteoroid impact every two days. My calculations, on the other hand, which are based on the comet flux at the Earth and on the Moon's smaller size, indicated that some 300 small comets were hitting the lunar surface every day. This apparent disagreement had everyone convinced that the small comets could not possibly be real.

Alex Dessler received a large response regarding the problem these comets posed for the Moon. One came from Ralph Baldwin of the Oliver Machine Company in Grand Rapids, Michigan. He thought that the Moon should quite literally have been sand-blasted by these comets. But the small comets are loosely packed porous bodies, not rocks, and there is a world of difference in what happens when such bodies collide with the Moon.

Nearly everyone would make the same mistake, including Paul Davis of the Department of Earth and Space Sciences at the University of California at Los Angeles. Davis had calculated the infall of comets into our celestial companion, based on its area, which is one-thirteenth that of Earth's. He assumed that the Moon's tidal forces would break up some of the small comets at an altitude of seventy-five miles, but that many comets of greater internal strength would reach the Moon intact. Finally, he calculated the seismic response these small comets would pro-duce upon impact from the disturbances known objects had produced on the Moon—the Saturn IV boosters and lunar modules that were sent crashing into the Moon following each Apollo mission. He concluded that every small comet hitting the Moon should have been easily detected by the lunar seismome-ters. Davis said, given the high impact rates of the small comets, the Moon should be in a constant state of vibration. It should appear, he said, "to ring like a bell."

Davis was essentially correct. If you treat a small comet like a rock, then the very sensitive seismometers that were placed on the Moon by the astronauts should have picked up the tremors created by the impact of these comets on the lunar surface. If you treat them like rocks then every comet that hits the Moon should have shown up on the seismometers, no matter where you placed these instruments. If these objects hit the Moon like hundred-ton rocks and the seismometers detected every one, then there would be more than 100,000 events due to these impacts on the Moon every year. But the lunar seismometers never picked up that many events or signals of any kind. There were less than 2,000 events a year. If you treat the small comets like rocks, then clearly that kind of discrepancy would end the issue of small comets.

Of all the Comments we received, the issue of cometary impacts on the Moon was the most difficult problem to work through. Unfortunately, I do not think many people understood either the Comments or my Reply, because everyone still seems to make the same mistake. The general complaint is that these large snowballs would have the same effect on the Moon as that of large rocks. They would produce the same craters and the same seismic waves. But these comets are not the same as rocks.

The problem can be stated in more familiar terms. Say I throw a one-pound snowball as hard as I can at my car. I then take a one-pound rock and throw it at my car with the same speed. It may come as a surprise, but the results are not the same. And they are not the same on the Moon either. I have a hundred-ton rock and a hundred-ton loose snowball. They are going to hit the Moon at the same speed. But the result will not be the same. All the arguments that were made against the small comets were based on the assumption that it made no difference whether you hit the Moon with a rock, a snowball, or a pillow of goose feathers. But given the choice, I would rather throw a one-pound snowball against my car than a one-pound rock. Davis obviously would just as soon throw the rock.

So would Dessler. Though he served as editor over all the Comments and Replies during the small comets debate, I do not believe he ever really understood the topic. In September of 1986 Dessler wrote me, not on official *Geophysical Research Letters* stationery, but on Rice University letterhead. He did this, he said, to make absolutely clear that the views expressed in the letter were not to be regarded as editorial decisions, but as helpful advice from one colleague to another. He was concerned about the quality of my Replies. He could not understand how the impact of such a body would fail to leave a huge crater on the Moon. He urged me to consider not replying to the Comments or saying simply that I did not agree with them. He thought that if the comets turned out to be fantasy, these Replies would compound the damage to my reputation.

But Dessler was wrong about the small comets. A snowflake shot at the speed of a bullet is not going to crack open a rock. But

Dessler has always insisted on treating every impact in terms of a nuclear explosion. So I suppose that given the chance to throw either a snowball or a rock at the windshield of his car, half the time Dessler would choose a rock. Dessler was also wrong about the Replies. Everyone I spoke to thought that a lack of a Reply to a Comment would be regarded by the community as an admission of defeat. I could agree with Dessler on only one point. If the small comets turned out to be real, he wrote, no one would remember who my critics were, much less what they said.

Few people ever understood the essential difference between a stony meteoroid and a cometary snowball. But Yosio Nakamura and Jürgen Oberst, both of the Institute for Geophysics at the University of Texas at Austin, did. Nakamura was one of the scientists involved in operating the network of seismometers that had been placed on the Moon. So he and Oberst knew what these instruments could and could not do. They had worked with the data and had analyzed them carefully for many years. So, with the help of Stephen Clifford and Bruce Bills from the Lunar and Planetary Institute in Houston, they considered in their Comment several conceivable scenarios to explain why the lunar seismic network might have failed to detect the small comets.

They realized that the problem could be resolved if the comets were objects of sufficiently lightweight material. Such a low-density object would be unlikely to produce a large Moon tremor. They calculated that if tidal stresses dispersed the small comets some sixty miles above the Moon, then all that would remain of each object upon impact is a cloud of snow flakes and ice crystals scattered over four miles. Such an impact would not penetrate the lunar topsoil. And the resulting disturbance would be so slight as to be undetectable by the seismometers. But for this to occur the break-up of the comets would have to be complete and no piece could be larger than the average-sized ones used in their calculations.

One way to get around this problem would be to assume that the objects are nothing more than large, thin clouds of matter long before they even reach the Moon. But this then poses another problem. How would such thinly dispersed objects survive for such a long time, given the heat from the Sun, in interplanetary space?

Nakamura and Oberst also tried to account for the failure of the lunar seismometers by assuming that these objects hit the Moon very softly. But they could think of no mechanism that would put the brakes on these objects sufficiently. Finally, they wondered if I had perhaps not grossly overestimated the size of these objects, for this too could account for the lack of seismometer readings. But I did not.

Nakamura and Oberst really tried to see the problem my way. They realized there is an inherent difference between a soft, flimsy object and a rock. They were aware of that. They realized what the problem was. They would not throw the rock through the windshield of their car. But they could not get from that insight to the response of the lunar seismometers. They stopped just short of understanding what these small comets were and what effect they would have on the Moon.

Even before the small comets came along, there had been some unanswered questions about the number of objects hitting the Moon. The number of large objects impacting a square mile of the lunar surface should be about the same as the number of large objects hitting a square mile of the Earth. The area of the Moon is about thirteen times smaller than the Earth's. Therefore, neglecting the larger gravity pull of the Earth, the Moon should receive about thirteen times fewer objects than we do. The lunar seismometers were calibrated so that scientists could triangulate the readings from the various instruments and determine the location, as well as the size and number, of the impacts. The number of large impacts on the Moon should have been about one-thirteenth the number of objects that had been seen to fall as meteors into our atmosphere. But lunar seismic data indicated that far fewer large objects hit the Moon than that. The discrepancy was substantial. There were about a thousand times fewer Moon shocks.

But the reason for the discrepancy is simple. It can be traced to the insensitivity of the lunar seismic stations for detecting objects that are loosely assembled like the small comets. Frank Press, a celebrated physicist from MIT and the president of the National Academy of Sciences, had worked on the physics of lunar impacts. He, along with Nakamura and a host of others, led the effort to have seismometers placed on the Moon. They calibrated these instruments by looking at the seismic signature of everything from nuclear blasts in the Nevada desert to bullets shot into loose sand and sand bonded with epoxy glue. Every event that would occur on the Moon would be scaled from nuclear blasts or from bullets hitting sand. No one ever worked out what effect a bullet of snow would have on such a surface. A lightly packed snowball cannot hit a surface the way a rock does. It simply does not, even if they are going at the same speed.

My Reply to the Moon problem was quite straightforward. The small comets do impact the Moon. No disruption above the lunar surface is likely since the tidal forces of the Moon are considerably weaker than the binding strength of the small comets themselves. So what happens when these forty-foot, low-density objects smash into the lunar surface at six miles per second? The effect is quite unlike the explosive scenario that accompanies the impact of a high-density stony meteoroid. To understand what happens when a snowball, or a porous body, hits a surface at high speeds I had looked at the brilliant work of a Russian physicist named Yakov Zel'dovich. He had done work on what happens to materials, including porous materials like water-snow, under high temperatures and pressures, though he did not apply his work to the lunar surface.

The evidence of a large rock hitting the Moon is a big crater. This is a permanent record. Most of the energy from a rock just buries itself into the Moon. For a small comet the same energy comes in, but it is spread out over a wider area. It does not end up in the lunar surface even though the same amount of energy is involved. The snowflakes of a small comet are widely separated from one another. So when the comet hits the lunar surface the flakes end up crushing one another. The comet does not instantaneously transmit all of its energy into the lunar surface. Most of the energy produced is crushing energy. So the front half of the comet comes down and gets crushed. It heats up and turns to steam. More and more flakes pile into one another. The crushing energy produces heat. Water temperatures may reach several thousand degrees, while some of the water itself is broken down into hydrogen and oxygen atoms. But very little of the energy is transmitted to the surface. Most of the energy goes into turning the snowflakes into water and the water into a hot steam.

Then once the collision is over, the vapor cloud cools and expands rapidly, kicking up dust. But the fast moving water particles escape because the Moon's gravity is so slight.

The lunar surface suffers very little damage from such an impact, just some wear and tear. A small comet hitting the Moon will not develop as much pressure upon impact as a rock. The pressure at the area of contact between the cometary water and the lunar surface is relatively low, though I would not want to be standing under it when it hits. It would still crush and vaporize me. Where the lunar topsoil is relatively hard the mass of hot water will essentially bounce off the Moon leaving no crater. On weaker lunar surfaces a shallow dent may appear. The resulting crater would most likely be flat and shallow-bottomed and measure about forty feet in diameter.

When a rock comes down and hits the surface, it hits hard enough and with sufficient force that a shock goes out from it, like the shock wave from a supersonic airplane. On the ground we experience this shock as a sonic "boom." The shock wave is produced because the object is moving faster than the particles in front of it can get out of the way. A large wave is produced, much like the bow wave in water from a high-speed boat. A rock that strikes the lunar surface is like the supersonic plane striking the air in front of it. It creates a shock wave in the lunar surface that propagates outward from the impact. Of course, a rocky meteor is destroyed as it tries to plow itself through the Moon. But if a flimsy object like a small comet impacts the Moon then no large shock wave is produced. And if it is flimsy enough and very lightweight then no shock is produced at all.

The extent of the vibrations produced by a ball of water-snow hitting the lunar surface cannot be as great as a rock's. It cannot move the Moon up and down and shake the seismometers the way a rock does. But it can still jiggle them. The instruments should still see a large number of events. But there is another factor involved and that is the frequency of the disturbance moving outwards from the shock wave. The lunar seismometers were tuned to a frequency of one cycle per second or one hertz, which corresponds to one up-and-down motion of the Moon's surface each second. An object hitting the Moon and creating any other frequency will not be as noticeable to the seismometers. Only if two tuning forks are matched does the movement of one fork cause the other to oscillate. In this case one tuning fork is the seismometer and the other is the Moon's surface.

The lunar seismometers were sensitive to the slightest fluctuation. But those fluctuations had better be tuned to about one cycle per second. The frequency caused by an object hitting the Moon depends on the size of the impact area, and this includes the distance that a shock wave moves before it is spent. The larger the impact area, the lower the frequency of the seismic wave. A rocky meteor creates a large impact area from a large shock wave, which moves the lunar surface at about one cycle per second— just what the seismometers were tuned to sense.

The small comets, on the other hand, produce little or no shock waves. Thus, their impact area is small and the waves travelling from the impact area have a higher frequency, about a hundred cycles per second. So the waves produced by the small comets would have passed right by the seismometers. A seismometer tuned to a one cycle-per-second wave will not pick up a hundred cycle-per-second wave. The instrument will be able to detect such an event only if it happens nearby. And the prediction is that a few will hit near the instruments if you have 100,000 comets impacting the Moon each year. But those may well be among the 200 or so impacts, thought to have been rocks, that were recorded each year by the seismometers. So the instruments had not recorded the small comet impacts simply because they had been tuned to hard rock collisions, while the comets, wrote one reporter, "splattered into the Moon at Barry Manilow wavelengths."

The seismometers' lack of sensitivity to the impact of small comets accounts for the discrepancy in the low number of large objects detected on the Moon relative to the number of such objects that are seen falling into Earth's atmosphere. It is not a factor of two or three, or thirty, but a thousand times or more. The difference is enormous. But the existence of the small comets can account for it. The larger number of objects falling into Earth's atmosphere is determined from the number of bright meteor streaks due to Öpik's puzzling dustballs. These dustballs are assumed to be large conglomerations of dust. But if the dustballs are small comets, then the observed large number of meteor trails in the Earth's atmosphere will be in agreement with the infrequent lunar seismic disturbances.

Oberst nearly admitted it himself. In a letter to Dessler, he said that he had found our Reply "most interesting" and went on to commend our "effort to find a formulation for the problem of lunar cometary impacts and their seismic responses." He agreed that "one of the causes of the discrepancy is indeed that cometary meteoroids are underrepresented in seismic data." But he maintained that cometary meteoroids could be identified on seismic data if "their kinetic energy is sufficiently high."

Once I had finished all the computations and completed our Reply I was personally embarrassed to learn that a pair of researchers had already conducted a detailed study of the impact of low-density materials on planetary surfaces. And they had done a much better job of it than I. John O'Keefe and Thomas Ahrens of the Seismological Laboratory at the California Institute of Technology in Pasadena had compared the effects of solid ice, silicate meteoroids, and porous snowball-like objects hitting solid surfaces at high velocities. Their computer simulations had shown that low-density objects would sometimes produce "a flat-floored crater with a central peak," but that "for very low density impactors," the excavation depth was "near zero."

Their work had been published in 1982 and I had missed it. But Dessler had given me just four weeks to come up with a solution to a very difficult physical problem, and I did not have the time to search all the literature before completing my Reply. If I had found their work beforehand, I might not have had to work it out independently. But I was pleased to learn that my schoolboy calculations had yielded the same results as their very thorough simulation with a large computer.

I think we have seen the effects of the impact of small comets on the Moon. On the evening of April 21, 1972, Thomas Mattingly, the command module pilot on the Apollo 16 mission, reported seeing a bright flash on the Moon's far side. No seismic signals were detected. So the scientists responsible for operating the lunar seismic network are well aware that their instruments did not pick every significant lunar event. If a rock had hit the

Moon to produce this flash, the seismometers would have detected it.

It was not the first time—nor the last time—that a bright flash or strange glow appeared on the Moon. More than 1,400 such events have been recorded by both amateur and professional astronomers and their reports date back more than 400 years. These events are known as Lunar Transient Phenomena. No one knows exactly what causes them, but many observers interpret what they see in terms of volcanic activity, despite the fact that the Moon is supposed to be a dead body.

I think many of the bright flashes and strange glows seen on the Moon can be attributed to the impact of small comets. As a comet slams into the lunar surface its kinetic energy is transformed into heat. The heat turns the water-snow into liquid water and then into a hot gas of water, hydrogen and oxygen. This is a very strange gas because it is under a very high pressure. Just as it blows apart, the hot gas glows. This is just momentary, but it must glow. The gas is so compressed and has so much energy that it glows just like the filament of a light bulb. To an observer on Earth looking up at the Moon with a telescope, it would appear as a brief, dull glow or red light. If that person happened to be looking at the mountains for volcanic action, they would see this glow and likely exclaim, "Volcanoes!"

Early in 1989 Sigwarth came running into my office with a copy of an article from the latest issue of *Icarus*. It was entitled "Photographic Evidence of a Short Duration: Strong Flash from the Surface of the Moon." It was fabulous. On the night of May 23, 1985, astronomer George Kolovos and his colleagues at the University of Thessaloniki in Greece had taken a series of photographs of the Moon with a small portable refractor telescope. In one frame a striking "bright spot" appears that is absent from both the preceding and following frames. Kodak Laboratories in Athens found no fault in the grain structure of the film. And the brightness of the spot ruled out the possibility of pre-exposure on that part of the negative.

The flash lasted less than sixteen seconds and illuminated the region around the crater named Proclus C. What is most striking about the photograph is that the shape of the bright spot is confined by the irregularities of the surrounding lunar features.

This strongly suggests that the event is real. And it seems, from an examination of the illumination on the inner edge of a crater some thirty miles from the event, that the flash took place "from a point slightly above the surface of the Moon."

The astronomers did their best to explain the mysterious flash. They ruled out the possibility of reflections, matter-antimatter annihilations, as well as volcanic eruptions given the absence of any sign of plumes in the subsequent frames. They also ruled out a meteor impact because no great cloud of dust was seen. Finally, they concluded that the flash had occurred after the lunar surface had been stressed, probably due to the quickly rising temperatures caused by the rising Sun. An intense stress, they believed, might trigger a cracking of the lunar surface, releasing gases and electrical discharges, and producing a momentary bright flash on the lunar surface. They calculated that the lunar surface would have to be heated to a temperature of about 3000°F to produce this flash. This, of course, is exactly what we had predicted would happen to the gases from the impact of a small comet.

There is yet a more permanent exhibit of the impact of small comets on the Moon and that is the zodiacal dust cloud. This is a cloud of dust and gas in the plane of the planets that extends from about Venus to the asteroid belt, just past the orbit of Mars. In reflected sunlight this cloud is faintly visible from Earth and is known as the zodiacal light. It appears as a diffuse glow in the western sky after twilight and in the east before dawn. Scientists have long debated the origins as well as the stability of this dust cloud over time. The supply rate has been calculated to be several tons a second. That means you have to pump in several tons of dust per second in order to maintain the known dust stream between the planets. Some scientists believe that the dust is kept in its orbits through shepherding by the Earth and the other planets in the inner solar system.

But there is no need for any fancy shepherding effect to get the dust ring in the inner solar system. Some dust must come from the disruption of small comets that are destroyed by the heat of the Sun as they travel in interplanetary space. And some dust has to come off the Moon when it is hit by small comets. The dust is blown up off the Moon with enough speed to be injected into these orbits. My calculations show that if small comets are hitting the Moon and the dust is being blown off of it, then the orbits should extend from about Venus to Mars in a sort of disk. So it seems likely that the small comets are contributing at least some material to this giant cloud of interplanetary dust.

I am not the only person to think so. A Uruguayan astronomer named Julio Fernández, who has apparently never heard of my small comets, recently proposed in a journal of planetology called *Earth, Moon, and Planets* that disintegrating comets may be supplying the zodiacal dust cloud with several tons of dust per second. He includes among them the known population of short-period large comets, as well as a population of "unobservable small ones." He is aware of the scientific consensus that comets with diameters under about a half mile are either scarce or non-existent. But, says Fernandez, rather than attribute this scarcity "to the actual absence of small comets in the inner planetary region, we can argue that they do exist but are very difficult to detect due to their faintness...." Amen.

One last problem remained. If 300 small comets do hit the Moon every day, where, some people wanted to know, was all the water on the Moon? One of the strongest conclusions from the study of the lunar rocks brought back by the astronauts is that the Moon is remarkably dry. But this problem is very simply resolved. The lunar gravity is such that practically all the water vapor from the impact of small comets simply flies off. But some of the water molecules may wander around and eventually condense in the crevices near the poles. A number of scientists, such as James Arnold, a chemist at the University of California in San Diego, have argued that water and ice may be trapped in the permanently shadowed regions near the lunar poles. The small comets may be a source of that ice.

Because most of the water flies off, a huge envelope of water should surround the Moon. We know the Earth has its own tenuous shroud of hydrogen gas lying beyond our atmosphere. The impact of the small comets onto the Moon's surface gives our companion a hydrogen shroud also. The cometary water vapor is hot enough that it escapes the Moon's weak gravity and moves outwards into interplanetary space. After travelling for about a day from the Moon these water molecules are broken apart by solar radiation into hydrogen and oxygen. These molecules will have travelled, on the average, over half a million miles. This means that our planet is sitting in a very tenuous lunar atmosphere, an atmosphere of water, hydrogen, and oxygen. Unlike the known comets, which expend their own water supply, the Moon merely releases the water deposited by the small comets that collide with it. Yet for a space traveler viewing the Earth and its companion from a distant vantage point, say Venus, the Moon should resemble a faint comet.

On March 7, 1971, several bursts of water vapor were observed on the lunar surface by a set of instruments, ion detectors actually, that had been left on the Moon by the Apollo 12 and Apollo 14 astronauts. The detectors had been built and the experiment designed by John Freeman and three colleagues at Rice University. Freeman, by the way, had gotten his Ph.D. at the University of Iowa. He was looking for ionized molecules and atoms in an attempt to measure the Moon's very weak atmosphere. Ionization occurs when an atom or molecule is stripped of one or more electrons. This stripping can occur by exposure to solar ultraviolet radiation or through collisions with other particles.

Shortly after analyzing some of the data, Freeman ran across an impulsive burst of water ions that had been detected by both the Apollo 12 and Apollo 14 instruments. It started a pandemonium. Everyone thought there was water coming up from the Moon. The existence of water in the Moon is a question of great importance. One of the problems with establishing a future manned base on the Moon, for example, is the lack of water. But if the Moon had its own water, the astronauts could simply pump it out.

In order to prove that the water had come from the lunar interior, Freeman had to eliminate all other possible sources for the water burst. So he compared the features of this event to a variety of known events. The only other occurrence that had produced rates as high as that of the 7th of March was the engine exhaust gases released during the takeoff from the Moon of the

Apollo 14 lunar module. In that case, the water vapor was present for only two minutes, while on the 7th it had lasted, intermittently, about fourteen hours. Besides, the lunar module had left the Moon a month prior to the mysterious event of the 7th.

Freeman also considered the possibility that his data actually represented a measure of the dumping of nearly ninety-five pounds of waste water from the Apollo 14 command module in lunar orbit a month prior to the incident. Urine samples were collected and their contents were compared to the water burst of the 7th. But Freeman ruled out this possibility as well. The event could not have occurred so long after the command module had left the Moon. Nor could the quantity of water produced by the waste dump have been as large as that observed on the 7th—anywhere from half a ton to eleven tons, depending on how far the instrument was from the source of water itself.

A number of other possibilities were considered and rejected, including a comet as the source of the water burst. Freeman ruled out the impact of a comet for several reasons. March 7th was not the date of any known meteor shower or the date when the Moon was near any known old comet orbits. Comets are also thought to contain significant quantities of other materials such as carbon dioxide, and these were not detected by the sensors. Finally, the notion of a cometary hit was "firmly ruled out," Freeman wrote, "by the lack of any seismic activity characteristic of such an impact."

Of course, no one at that time knew of the existence of small comets.

Chapter 19

Mars and the Outer Planets

The Moon and the Earth are not the only targets in this cosmic shooting gallery. Much of the rest of the solar system is likely to feel the impact of the small comets as well, though nowhere can we expect their presence to cause such a bounty—oceans, atmosphere and possibly life—as on the Earth.

I expect very few small comets to survive inside of Earth's orbit where they would be destroyed by the Sun's heat. No small comets will make it to Mercury, few will make it to Venus. But it is quite a different story for the remainder of the solar system, especially Mars, Jupiter, Saturn, Uranus, and Neptune. The small comets have to be there. Their numbers are large, their effects considerable.

The flux of the small comets in the vicinity of Mars is about the same as it is at the Earth. But Mars is about half the size of the Earth, its gravity is a little more than a third, and its mass is just a tenth that of Earth's. So my estimate of the rate at which the small comets impact the Martian atmosphere is approximately one every twenty-five seconds. That is quite a bit less than the twenty per minute for the Earth, but it is still enough to bring a large amount of water to Mars. The small comets would deposit about three feet of water on Mars every 400,000 years. So, everyone wanted to know, where is all the water on Mars?

Pictures of Mars taken during the United States Viking mission show what looks like a desolate red desert. Nowhere is there any running water. But the Martian surface displays remarkable evidence that water once flowed on it in great quantities. The terrain is rippled with long channels complete with branching tributaries, teardrop-shaped islands, and braided silt deposits. "These networks resemble terrestrial stream networks that are the result of collected rainfall," according to a thorough analysis of these features by a team of U.S. Geological Survey scientists headed by Harold Masursky.

Masursky determined the age of these features using the standard method of crater count. The more heavily cratered an area, the older it is thought to be, because meteor impacts are assumed to happen at more or less the same rate over most of the lifetimes of the planets. Therefore, thought Masursky, since many Martian channels show so few craters, they must be quite young. These stream channels, he concluded, "may have formed during relatively short periods of time at several different times." There must then have been episodes in the history of the solar system when the atmospheric pressure and the temperature of Mars were great enough for liquid water to flow freely.

The atmosphere of Mars today is too thin for liquid water to form on its surface. But the planet does have ice caps at both poles that advance and recede with changes in the Martian seasons. These ice caps were originally thought to be largely dry ice, or frozen carbon dioxide, with very little water. So they certainly could not be a reservoir for an ocean or a fraction of an ocean of water pumped in over time. But scientists have come to believe that there may actually be more and more water in those polar caps than originally thought. The evidence now suggests, according to Hugh Kieffer of the U.S. Geological Survey and his colleagues, that, "the water deposits are fairly thick" in the polar area.

What happens to a planetary atmosphere if it is all water? No one knows really, but such a scenario was worked out years ago by Andrew Ingersoll of Caltech to explain the early water loss on Venus. I think the same scenario is applicable to Mars. If you have a water atmosphere, it will eventually get hot. It will be just like the greenhouse effect on Earth, but it will not be due to carbon dioxide. Water can also produce a greenhouse effect. So as comets come in and deposit more and more water into the atmosphere, it gets hotter and hotter. Eventually the polar caps melt, producing even more water. Mars has little gravity and does not keep its atmosphere as well as the Earth does. So eventually the Martian atmosphere would get so large and so hot that it would finally have to just blow off into space. There is a name for this among planetary scientists. It is called hydrodynamic blow-off.

Here then is what may be happening on Mars. The small comets come in and penetrate to a Martian altitude of about twenty-five miles. They vaporize. There is enough air in the weak Martian atmosphere to vaporize the comets. They do not hit the surface—not the ones that are hundred tons—just as they fail to hit the surface here on Earth. So they enter the atmosphere, vaporize and then condense as ice at the polar caps. But eventually there is so much water and ice on Mars that the incoming water vapor cannot be collected out of the atmosphere. Then you have the greenhouse effect. Solar energy is absorbed and trapped in the atmosphere. It gets hotter. Once the surface gets to the melting point of ice, you get more water melting from the polar caps. Water begins to flow in the channels on Mars. This is the Martian springtime. It must be an amazing scene.

But how long does it last? What time passes between spring-times? Will there be violent thunderstorms? Can simple forms of life develop during the springtimes and hibernate during the long droughts that may last for millions of years? We do not know. We have never seen Mars in such a warming stage. But then again we have been keeping tabs on the planet for but a tiny fraction of its four-billion-plus-years' history. Eventually, how-ever, its atmosphere would grow so large and so hot that a great loss of water into space by hydrodynamic blow-off would occur. Within a short period of time all the water would be gone and the process would start all over again.

Such a view is not inconsistent with some models of a cyclical Martian climate. Some planetary scientists believe that Mars has experienced such periodic warmings in the past. Direct evidence of such a warmer period, complete with liquid water, has been reported by a team of scientists headed by Ted Roush at the NASA Ames Research Center. The evidence lies in the planet's dust, which periodically rises into huge dust storms that engulf the Martian surface for months at a time. Roush has found signs of volatiles, substances that evaporate at relatively low temperatures, in spectroscopic observations of the Martian dust. These dust grains, he said, may act as permanent reservoirs for carbonates, sulfates, and water.

Fraser Fanale, a planetary scientist at the University of Hawaii in Manoa, has another idea to dispose of, or hide, a large amount of water on Mars. It did not occur to me because I had read the papers about the polar caps and it seemed clear that there was no ocean of water there. Of course, Fanale was not trying to account for the small comets. He was involved in another fierce argument about whether or not there is more water on Mars than we think. People are unhappy with him because he also is water prone.

Fanale's idea is that the water condenses on top of the polar caps, just like the frost that collects in a refrigerator. Gradually it builds up, putting more and more weight on the polar caps, until the ice slides under the surface dust. He suspects there may be as much as a mile of water under the polar caps. Others have had similar thoughts. Back in 1980, Stan Zisk of MIT's Haystack Observatory reported finding an area of Mars which appeared to have liquid water some twenty to forty inches below the surface. His evidence came from a set of Earth-based radar measurements of the Martian surface carried out a decade earlier by the Jet Propulsion Laboratory's Goldstone Facility.

The notion of subsurface water on Mars is a clever idea. Some future exploration of the Martian surface may one day provide us with the deciding evidence. But until then current wisdom alone is not sufficient to eliminate the possibility of a large influx of small comets on Mars.

Also certain to feel the impact of these objects are the moons of Mars and those of the outer planets. Phobos, a Martian moon, shows extensive alteration by liquids and is thought to conceal ice below its surface. Jupiter's Ganymede, the largest moon in the solar system, resembles a huge cratered ball of ice; Callisto is also covered with ice. Most of Saturn's moons are relatively small but they too are composed of rock and ice. And many of the natural satellites of Uranus are also ice-clad. Where did all this ice come from? Quite likely from the infall of small comets.

But determining the effects of small comet impacts on planetary moons requires more than fleeting thought. The farther a planet's position from the Sun, the less the impact speed of small comets on its moons. On the other hand, the closer the moon is to a giant planet, the higher the speed of cometary impacts because of the boost in speed due to gravity. For our Moon, the impact speed is such that cometary water is blown off into space. But the middle and outer moons of the giant planets should retain the water from the impact of small comets, particularly if the comets penetrate into a snowy or icy moonscape.

One such candidate is Neptune's Triton, which is now regarded as the coldest body in the solar system. Its exotic surface, an incredible icescape of what appears to be frozen nitrogen oceans, was revealed during the Voyager 2 flyby in 1989. Some think the large shapes on the surface are made of water ice since ices of methane and nitrogen could not be used to build the large vertical structures that were observed. Most intriguing are the volcanic-like geysers that rise out of the icescape and are swept leeward by Triton's atmosphere, looking like chimneys trailing smoke. Some believe the geysers are produced when dark material on the moon's surface absorbs sunlight and heats the nitrogen which then escapes through cracks in an ice layer about eight feet thick. But how can this be? The Sun's heat is about one-thousandth that at Earth and the dark features have been there a long time. The heating of frozen nitrogen, however, could easily be produced by the impact of a small comet. But no one has mentioned this obvious possibility.

The origin of planetary rings poses another puzzle for planetary scientists. While such ring systems are absent from the smaller inner planets, all the giant outer planets have them. There is a single, narrow wispy ring that extends out from Jupiter almost 186,000 miles. Saturn's rings, of course, are broad and magnificent. Uranus has ten dark narrow rings and Neptune has five faint rings. These rings are all composed of icy particles. Where did all the ice come from? Planetary scientists now reject the notion that these rings are composed of material left over from the time the planets were forming. And some are beginning to realize that these ring systems require a continuous supply of new material. I think the small comets may be one such source. Perhaps some small comets are colliding with rocks and tiny moons orbiting in the rings and leaving some of their water ice and mantle material in orbit to continually rejuvenate the rings.

I believe that the fingerprint of the presence of the small comets is actually visible as dark spokes in Saturn's rings. It seems that dark, radial, wedge-shaped spokes are continuously appearing and disappearing in Saturn's B ring. Fresh spokes have been observed to form within twelve minutes across part of the rings. A small comet is fast enough to do this. So these spokes may well be the tracks of small comets moving across the rings as they fall into Saturn's atmosphere.

The influx of small comets into the giant outer planets—though necessarily just rough estimates based on their population at the Earth—is in all probability absolutely tremendous. Take Jupiter, for example. It is about three hundred times more massive than the Earth. And because it is bigger, its gravitational field is larger. Consequently Jupiter probably sucks in some 500 small comets per second. So over a period of four billion years Jupiter acquires the equivalent of one Earth mass in cometary water. Such an infall of material will produce a lot of heat.

I was shocked when I first calculated the heat produced by these small comets at the outer planets. I thought I had stumbled onto something important and I mentioned it briefly in our original set of papers in *Geophysical Research Letters*. Later Donald Morris at the Lawrence Berkeley Laboratory brought up the subject of planetary heating in an early version of his Comment on the small comets. It was an amusing situation. He first used it as a way of eliminating small comets. Then, when the heat from the small comet impacts was shown to be reasonable, Morris claimed the idea as his own. We ended up having quite an argument about it. Alex Dessler took his usual route and eliminated these tantalizing suggestions for planetary heating from both the published Comment and Reply. It was another blow for scientific inquiry.

The topic is an important one, however. The Sun is twenty-five times dimmer at Jupiter than at Earth. But Jupiter radiates more heat than it receives from the Sun, about ten percent more, in fact. Its neighbor, Saturn, radiates a great deal more heat than it receives from the Sun. And Neptune also radiates more heat than expected. Only Uranus, among the giant outer planets, does not appear to generate excess heat, although earlier reports claimed that it did. The determination of excess heat is difficult.

Current wisdom interprets this excess heat emission in terms of an internal heat source. In the case of Jupiter, the excess heat is thought to be the energy released by slow gravitational shrinkage. This excess heat is one of the major features in the modeling of the history of the large planets and what lies in their deep interiors. But it turns out that the 500 comets that slam into Jupiter every second are enough to account for at least some, if not all, of the excess heat coming from Jupiter without resorting to internal heat sources.

The energy flux from the comets is big enough, within a factor of two or three, to account for the difference in thermal energy production of not only Jupiter, but Saturn and Neptune as well. Uranus remains an anomaly for everyone. So if this excess energy is coming from the small comets rather than internal heat sources, then many of our accepted ideas on the evolution of the large planets and their interiors are wrong. Our current ideas regarding the history and nature of the entire solar system may have to be revised.

But there is no need to worry about the Earth. The impacts of these small comets provide little heat for our planet. The Sun produces by far the largest share of the energy on the Earth. Additional heat sources are radioactive material in Earth's interior and tides produced by the Moon. Besides, the Earth is a small planet. Its gravitational field is much smaller than those of the giant outer planets, so the small comets are not sucked into the atmosphere as efficiently or at such high speeds. The small comets merely supplied us with most of the basic material for our bodies, as well as the benign stewpot for the evolution of life itself.

The Earth is pelted by about 10 million small comets a year. Mars receives a smaller annual share, less than a million and a half. Giant Jupiter is bombarded by an incredible 16 billion small comets each year. Smaller Saturn gets by with about four billion, while Uranus totals 260 million and Neptune with 300 million. Tiny Pluto is blessed with a sprinkle of only 500 thousand small comets a year. That brings the total annual budget of small comets for the solar system to something over 20 billion. That is a lot of comets to store somewhere over the age of the solar system and I quickly began to wonder where all these small comets could possibly come from. That was the next subject we had to tackle. It would not be an easy task.

Chapter 20

Where the Small Comets Come From

With so many small comets colliding with the planets, it seemed that any attempt to trace their origin would spell their demise. Many people thought the debate would come to an end because there was just no possible source for ten million small comets plummeting to Earth alone each year for more than four billion years. Where could they all come from? Where could we store so many objects? Unless we could find a likely reservoir, I realized that the small comets could not possibly exist. The basic problem—and one I did not have solved at the time my two original papers were published—is where did these objects come from?

We had to find a believable source for this massive collection of small comets; we had to find a way of mathematically sticking the toothpaste back into the tube. There were several possibilities. The first that came to mind was a reservoir of small comets in the space between the planets. This is a convenient solution, but unlikely. Between pulls caused by the gravity of the planets and destruction from heating by the Sun, it is just impossible to store so many comets in the neighborhood of the planets over the age of the solar system.

The next possibility is that we live in a very special time in the nearly five billion year history of the solar system. What would be special about it is that somewhere in the solar system a very large comet exploded and made a lot of little comets. It would be a unique comet in unique times. The comet would need to be a million times larger than Halley's Comet to get the number of small comets we now see streaming by the Earth. But I worry about all the dust that would come from such a comet. The small comets have very little dust. So while I cannot say that this did not happen, I think the chance is very slim.

For a time we thought the most obvious supply for our large population of small comets would be the Oort cloud. This is the home of more than a trillion large comets and one of the best advertised features of our solar system. It was named the Oort cloud after the distinguished Dutch astronomer Jan Oort who first proposed its existence in 1950. This vast cloud of comets is located at a very great distance from the Sun yet is still loosely bound by the Sun's gravity. Its outer edge is thought to be located at a distance equivalent to 100,000 trips from the Earth to the Sun. Sunlight takes only about eight minutes to reach the Earth but it takes almost one year to travel to the edge of this distant comet cloud.

No one has ever seen the Oort cloud. Its existence has been inferred from the occasional large comets that we see wandering into the inner solar system. Oort showed that if you trace the paths of these comets back before their encounters with the planets, you can find the region from which these comets come. He showed that these comets were uniformly distributed around the Sun in a very distant spherical cloud. The objects that make up the cloud are moving in all directions like a bunch of hovering bees.

The Oort cloud, because it extends out a huge distance, is weakly bound to the Sun and therefore subject to disruption by passing stars and gravity tides from the galaxy. It is believed that the large comets we see have been scattered from this cloud and sent plummeting into the inner solar system. There are new large comets coming into the solar system all the time. Most are not as dramatic as Halley's. Some have orbits that bring them into view only once during our lifetime, during all of written history, or longer. Others will have smaller orbits and will pass by more often. All it takes is a little tug out in the distant Oort cloud and the comets come hurtling in.

If the small comets did come from this Oort cloud, our storage problem would be solved. But such a solution fails to explain the motion of the small comets. We know from the movement of those little fly specks in our images that the small comets circle the Sun in the same direction and in about the same orbital planes as the planets. But an arrival of objects from the Oort cloud would have a greater variety of orbits. Some of the objects would move in directions opposite to the Earth's motion and their orbits would not all be near the orbital planes of the planets. So if the small comets did indeed come from the Oort cloud—or from any passing interstellar cloud from another part of the galaxy, for that matter—you would see them coming in from other directions and at different velocities. But this is not what happens.

So the Oort cloud is not the reservoir that gave birth to the small comets. But if you go back to Oort's paper, published in the *Bulletin of the Astronomical Institutes of the Netherlands*, he points out that, if the cloud was there at the birth of the solar system, by now passing stars and galactic tides would have wiped it clean. It would not have lasted nearly five billion years. All of the comets in the cloud would have been stripped out of it. But we know the cloud is there, so it has to have a supply of large comets. Something must be feeding it.

Most work on the subject just ignores the fact that you have to supply new comets to this cloud. Oort proposed that the cloud was being fed by an inner disk of comets, but most people do not refer to it. When the solar system was formed, the Earth and the planets condensed from dust and rock, and out beyond the planets the comets formed from left-over icy grains and dust in a great disk aligned with the orbital planes of the planets. So a big reservoir of large comets was formed, circling just out beyond the planets. They are closer than the Oort cloud itself, and more tightly bound to the Sun's gravity, so a passing star would not jiggle them as much. This is a more secure storage place. But every once in a while a star might pass very near or through this inner disk and scatter some comets. Some would feed the outlying Oort cloud, some would be lost into interstellar space, and others would plummet into the inner solar system.

In 1981 Michigan State University astrophysicist Jack Hills confirmed that the Oort cloud itself could only be supplied by this inner region of comets, the Oort disk, and then showed, contrary to popular opinion, that many more large comets actually come from this inner region than the Oort cloud itself. He calculated that the total number of comets which have entered the planetary system directly from this Oort disk is about twenty-four times the number that have come from the Oort cloud itself. The total mass of this inner disk, says Hills, may be a hundred times greater than the Oort cloud.

This storage ring beyond the planets is now used by many people to stow away large comets. It is not my idea. I just stuck some small comets in with the large ones. I thought it would be reasonable to place the large comets near the inner edge of this great disk and the small comets in the outer portions of the disk. My reasoning is quite simple. Some scientists believe that the dust in the early solar nebula was concentrated on the inside of the nebula, while the ices or volatiles were located primarily on its outer rim. It would seem, therefore, that the present Oort disk would be populated with large dusty comets at its sunward edge, while small comets made almost exclusively of volatiles such as water would reside further out.

I think that this relatively massive disk of matter beyond the planets left over from the formation of the solar system nearly five billion years ago is the only possible source for the small comets. If somebody could show that there is no disk out there, then there would be no small comets either. There is no other way of supplying the small comets. Of course, we cannot see this disk within our own solar system. But we know that such features exist in the cosmos. Recently, astronomers have been able to obtain images of other stars with vast disks like this.

The Infrared Astronomy Satellite, a joint Anglo-Dutch-American endeavor, found several stars surrounded by such disks in 1984, the most famous one of which is the great disk of Beta Pictoris as seen edge on. Soon afterwards, Bradford Smith of the University of Arizona and Richard Terrile of the Jet Propulsion Laboratory were able to photograph this enormous disk of matter in visible light. They did so by blocking out the light from its hot bright sun so as not to overwhelm the more feeble light reflected off the disk. It is a beautiful picture, a first peek at planet formation.

We know much more about the birth, life, and death of stars than we know about planets. Because we can point our telescopes at the billions of stars of all ages and types, we can determine what happens to a star from the moment of birth onward. But we cannot do this with planets because the only planetary system that we can observe directly is our own. We have no pictures of planetary systems younger or older than ours, so we must guess how the planets were born and how they age. The recent detection of large disks of gas and dust around newborn stars, such as the one around Beta Pictoris, is a giant step toward that elusive goal.

The disk around the star Beta Pictoris extends out a distance equivalent to almost a hundred trips from Jupiter to the Sun. Planets are thought to be forming very near the starward edge of the disk. But while this may be where most of its material lies, there is considerable matter far beyond it. The disk actually thickens as the distance from the central star increases. Smith and Terrile believe the disk may be partly or entirely made of small grains of silicates, carbon-based matter, and ices, primarily water-ice.

Beta Pictoris is in its early formative days, but its features are applicable to our solar system because there is no way to strip off such a large disk of matter over some five billion years. It is too tightly bound to its star to have other stars carry it off. If these observations are applicable to our solar system—and there is no reason to believe that Beta Pictoris is unique—then it is likely a disk of similar matter lies somewhere out beyond our most distant planets, Neptune and Pluto.

Astronomers are beginning to realize that such a storage bay of comets just beyond the planets is not only likely but necessary. Martin Duncan of the Lick Observatory and Thomas Quinn and Scott Tremaine, both of the University of Toronto, recently found that the most likely origins of large comets moving near the orbital planes of the planets is a disk of such objects near the orbit of Neptune. This is less than a thousandth the distance to the outer edge of the spherical Oort cloud.

Duncan and his colleagues came to this conclusion after extensive calculations using months of computer time. They were looking for a way to explain why the hundred or so large, short-period comets all lie primarily in the planes of the planets and orbit the Sun in the same direction as the planets. The orbits of this special class of large comets are remarkably similar to the small comets. They found that even a long game of "gravitational pinball" with the giant outer planets could not herd large comets into these orbits if they came from the outer Oort cloud. So they went back to the view offered in 1951 by Gerard Kuiper. He had proposed that cometary material left over from the formation of the solar system lies in a disk just beyond Neptune's orbit. But Duncan and his co-workers faced one bothersome point in their proposal: They could not understand what shakes comets out of their storage ring and sends them plummeting into the inner solar system.

But once the existence of an inner cometary disk is established, the solution to this puzzle and to the origin of the small comets is largely a matter of Newton's laws. It has to do with the motion of bodies moving in each other's gravity. But while it is easy to describe a pencil, or a plate of food, falling on the floor, when we are talking about the solar system, we are talking about many bodies moving out there, pulling on each other all at once, and it is no longer so easy to describe. It is, in fact, enough to give anyone a headache.

Just because there is a disk of comets out there beyond the planets does not mean that any of them are ever going to reach the Earth. Left on their own, the comets will remain in the disk until the end of time. But a passing star might bring them in. Such a star would have to run a nearly parallel course through the disk, however, and that is an unlikely occurrence. Otherwise the brief passage of a star through the disk will only produce a drizzle of comets. Not enough of them will reach the Earth to account for the numbers of small comets seen by the Dynamics Explorer satellite. This is a severe problem. If a large object is not there a long time to scatter the small comets, and the small comets do not reach Earth, then we are not here either. All the volatiles needed for the origin of life on Earth are out there.

I think the most likely candidate for the task is a large undetected object the size of a planet. This planet has to somehow scatter ten million small comets into the Earth's atmosphere each year. Many more than this must reach the Earth's orbit without colliding with our planet. This is a huge requirement. What happens if a large planet plows through this rotating disk of comets? It scatters some of them at just the right angle to send them into the inner solar system and the Earth. Those that do not strike the Earth, the Sun, or the other planets, will return to the distant disk and eventually travel back to the planets again. That is how you get the small comets to Earth. But this number of small comets is too low. It will not work. And the reason the number at Earth is too low is that it takes a long time for the small comets to zoom clear out into the disk and back. They do not orbit by the Earth often enough to provide the flux of small comets I need. So the scattering of small comets in the inner Oort disk by a distant unknown planet does not by itself account for the large flux of small comets at Earth.

Something more is needed. What is needed is some way to decrease the orbital times, and thus the orbital distances, of the small comets. If the orbital times decrease then the number of times that each comet crosses the Earth's orbit increases. In this way the same number of comets gives you a far greater flux at Earth than you would have otherwise. This is where the work of physicists Edgar Everhart and Julio Fernández came in. Both used computers to see what would happen to comets in long period orbits when they were disturbed by the giant outer planets. They were talking about large comets, of course, but their work applies just as well to small comets. They found that some comets are destroyed by the Sun, others are lost to interstellar space, and some comets end up in orbits with short orbital times around the Sun. This happens because some of those that reach the inner solar system have to pass close by Jupiter and Saturn. The gravities of these planets scatter the comets into short period orbits that swing out only as far as Jupiter and Saturn. This allows them to go by the Earth several thousand times, rather than once during the same period of time if they had orbits that took them back to where they came from, 300 times further away.

The giant outer planets are responsible for this kind of amplifier effect. It turns out that about one in a hundred comets ends up with a short period orbit. These comets move in the same direction as the planets around the Sun and stay near the orbital plane of the planets, just like the small comets. To give the giant outer planets sufficient time to capture a large enough number of small comets to get the proper infall rate at Earth, the comets would have to have a lifetime greater than 100,000 years. Fortunately, that lifetime is within the range I had estimated for the small comets given the protection of a carbon-based mantle.

This is an incredible series of events. First you need a large storage disk of comets and a large, errant planet. But this planet cannot have a circular orbit like the others, so you need a star as well that long ago passed close enough to the planet, but not so close as to destroy it or pull it away from the solar system. So now you have this large planet wandering through the disk and scattering some comets. Then you need the presence of the giant planets like Jupiter and Saturn to draw the comets into shorter period orbits. Only then do we get the right number of small comets to produce the oceans, a portion of the atmosphere, and, most likely, the organic material for life on Earth. It bothers me that this is such a unique system. But it is plausible.

Yet this set of special conditions makes life on planets around other stars much less likely than previously thought. So perhaps our lack of radio contact with extraterrestrial civilizations is due not to a quarantine for our barbaric social behavior or to the weakness of our radio telescopes, but because we are simply alone.

Chapter 21

Death of the Dinosaurs

In looking for the source of the small comets, we ended up explaining far more than anyone ever expected. Any way of herding small comets into the Earth's vicinity was likely to bring in large comets as well. But while the everyday showers of forty-foot comets may have brought the ingredients for life to this planet, a storm of mile-wide comets would most likely spell death and destruction for much of life on Earth. Suddenly many parts of the puzzle began falling into place, beginning with why the dinosaurs disappeared 65 million years ago.

Much of the groundwork for such a scenario was laid in 1980 and the following years by a team of four scientists that included the late Luis Alvarez, a physicist who won the Nobel Prize in 1968 for developing bubble chambers to investigate the structure of the nuclei of atoms; his son Walter Alvarez, a geologist at Berkeley; and Frank Asaro and Helen Michel, both nuclear chemists at the Lawrence Berkeley Laboratory. They had come to the conclusion that a mountain-sized object had hit the Earth 65 million years ago and caused the extinction of much plant and animal life. They had focused mostly on the extinction of microscopic marine animals, the phytoplankton and zooplankton, as well as some larger marine organisms. But the dinosaurs had died rather suddenly and at the same time as well.

Their startling proposal created considerable controversy. The very thought that our destiny on Earth could be affected by an extraterrestrial object conflicted sharply with current wisdom and the emotional issue of our isolation from the rest of the solar system. The intense debate that resulted forced scientists to seriously consider whether or not our destiny might be tied, at least in part, to extraterrestrial events of this kind. Since then science has come to accept this possibility, though reluctantly, and only to explain events of the past. But that such celestial intervention might be ongoing today is still, for science, quite a taboo subject.

Much of the evidence for the Alvarez idea involves an unusual layer of clay that separates two periods in geologic history, the Cretaceous and the Tertiary. This boundary not only contains the last burial ground of the dinosaurs but an unusually high concentration of iridium, a rare element that is, however, much more abundant in meteorites than in the Earth's crust. Walter Alvarez amazed everyone with his discovery that the bottom of this clay layer contained 300 times more iridium than the rest of the Earth's crust. This strongly pointed to an extraterrestrial event.

To explain this mystery the Alvarez team of researchers proposed the following scenario. An object measuring about six miles across struck the Earth 65 million years ago. The material from the explosion enveloped the planet in a blanket of dust. This dust fell onto the ground and into the ocean and produced this clay layer in the geologic record. The settling dust would also have cut out sunlight, halted photosynthesis, and caused much of life on the Earth to die of starvation. The lack of sunlight might also have chilled the Earth causing yet more deaths. But perhaps the most important killing mechanism was due to the initial shock and heating of the atmosphere produced by the expanding fireball. This fireball would have caused unusual chemical reactions in the atmosphere and produced a deadly acid rain.

Geologists have now found three periods of iridium enhancement in the geologic record. This one 65 million years ago, another 38 million years ago, and the last just 11 million years ago. These dates suggest that something deadly occurred with surprising regularity every 26 million years or so. A similar periodicity has appeared in the work of two University of Chicago paleontologists, David Raup and Jack Sepkoski. Their vast collection of observations of ancient marine life in rocks shows that eight major episodes of extinction have occurred in the past 250 million years, each one about 26 million years apart. Additional evidence for the extraterrestrial connection comes from the work of Michael Rampino and Richard Stothers, both of the NASA Goddard Institute for Space Studies in New York City. They show that large terrestrial craters have been produced at nearly the same intervals. Evidently something from space has been killing off life on the Earth with striking regularity.

The Alvarez idea that something crashed into the Earth 65 million years ago was a good one, but it did not fit all the facts. The recurrence of extinctions at other periods in time suggests the long term effect of *showers* of objects, such as comets, rather than a single object. It is difficult to believe that the Earth would get struck by one large rock on a regular time schedule. Besides, paleontologists examining the fossil record have noted that the periods of extinction did not occur suddenly but gradually; they took place over several million years. And within these several-million-year periods they have identified separate periods of extinction, a phenomenon they call stepwise extinctions. Only a shower of large objects over time could produce such an effect.

A shower of comets can occur only when a reservoir of comets is disturbed. The culprit can be either a star or a planet. The chances of an errant star crashing through the solar system and sending comets careening toward Earth on such a regular basis are slimmer than your chances of winning the Iowa lottery. But people have tried to prove it. Richard Muller, a physicist at the University of California at Berkeley, tried to account for 26-million-year extinctions with a solar companion called Nemesis.

Muller places Nemesis in an orbit around the Sun that takes 26 million years to complete. Every 26 million years it zooms through the Oort Cloud and produces a shower of comets that causes periodic extinctions on Earth. The problem is that Nemesis could not possibly survive in such an orbit. It barely would be within reach of the Sun's gravity. Most likely, it would get thrown out of the solar system by passing stars or gravity tides that roam through the galaxy. It would not give you many 26-million-year cycles. Its orbit is just too big.

An unseen planet nearer the Sun is a better bet. Daniel Whitmire, a physicist at the University of Southern Louisiana, realized this and created Planet X. I think he had the right idea, but the wrong orbit. He did his homework correctly without a lot of hocus-pocus. He thinks that just beyond the known planets is an as-yet-undetected planet about the size of the Earth. This is not so far away, about a hundred times the distance from the Sun to the Earth. That means it is wandering out there about where Pluto is. In about the same location or a little further out Whitmire has this belt of large comets. This is his view of the Oort disk.

Whitmire places Planet X at this distance so that it will not be easily stripped away. The planet will not wander away as a star passes by. He is safe on that score. But he has another problem and he deals with it very cleverly. When you shrink the orbit of the planet the period around its orbit becomes smaller, so you can have no 26-million-year period for an object this close to the Sun. The best you can do is only a thousand years. So what Whitmire does is tilt the orbit of Planet X up a little bit out of the orbital planes of the planets. This orbit will very slowly change its position relative to the comet belt because of small tugs caused by the gravity of the planets. Most of the time Planet X will not be in the comet belt and hence there is no storm of comets at Earth. The only time it does so is when the most distant part of the orbit is located within the comet storage disk and that happens every 26 million years.

I like Planet X but it will not do for the small comets. Whitmire's scenario does not allow me to shovel in enough comets to the Earth. First of all, his comet disk is not big enough. I need to extend it out a hundred times farther. I have to have a disk that is at least comparable to the one seen at Beta Pictoris. Actually, I need one that is larger. I think the pictures we have of the disk at Beta Pictoris are just not sensitive enough to reveal its full extent.

The other major drawback to Planet X is that some of our spacecraft have now reached these distances and have discovered no such planet out there. Pioneer 10, for instance, has now moved to destinations outside the positions of the planets. If an object the size of Earth was located out there, we would have seen the gravitational pull of that object on this spacecraft by now. But we have not. So the object is not there.

But Whitmire's idea is magnificent. All I did was revise it. I first made the planet larger to create a greater disturbance in the Oort disk. Then I extended the orbit of this planet, but not too far, so that at its greatest distance the planet is still within the firm grasp of the Sun. This larger orbit allows me to reach out into that zone where the small comets are. This also means that the planet need not have been detected yet. Mine is not a new idea. It is just Planet X with more mass and a larger orbit. But these differences require that the planet have a new name. For obvious reasons, I have called it the Dark Planet.

I think there exists a Dark Planet that moves around the Sun in a great looping orbit that shakes billions of comets loose from the Oort disk. It takes about 500,000 years for the planet to complete its orbit. At its closest, this orbit brings the Dark Planet near the orbits of Neptune and Pluto or just beyond; at its farthest, it is perhaps another 150 times farther out. Most of the time the Dark Planet plows the outer regions of the Oort disk where the small comets are stored. If it is large enough, about the size of Jupiter or Saturn say, then the planet will disturb enough small comets to explain their steady rain on Earth. But the great number of small comets in the disk also creates a slight tug on the Dark Planet that moves its orbit slowly so that every 26 million years the planet enters the inner edge of the Oort disk, a region rich in large comets. When this happens tens or hundreds of large, ten-billion-ton comets plummet into the Earth over a period of about 10,000 years. The consequences for the Earth are enormous.

The environmental havoc resulting from such an intense, repeated bombardment of Earth should cause the extinction of various species and leave its mark just about everywhere. Alvarez believed that the event 65 million years ago was the result of the impact of a single great asteroid. But the distribution of iridium from a single asteroid should be heaviest in the vicinity of its impact. A large number of impacts caused by a comet shower, on the other hand, should produce a fairly even salting of iridium over the Earth's surface, and this is what is observed. It also happens that the duration of such a cometary shower is about the same as the interval that geologists have estimated it took to deposit the iridium in the Cretaceous-Tertiary boundary.

Also expected at this geological boundary is carbon from the incomplete burning of the comets as they crashed through the atmosphere. This has been found by geologist Wendy Wolbach and her colleagues at the University of Chicago, though they interpret its presence in terms of global wildfires ignited by the impact of one enormous meteor. Shocked quartz and large craters are also expected from the impact of large dusty comets. This shocked quartz has also been found and a survey of the ages of large craters on Earth shows a periodicity of about 28 million years.

The case for a single impact is further weakened by the evidence for the stepwise extinction of species, a series of catastrophic events, during this period of time. But these step-wise extinctions are easily explained by the motion of the Dark Planet's orbit. Several successive orbits of the Dark Planet will cross the inner region of the Oort disk during each 26-million-year interval, since the planet's orbital period is short compared to the slow turning of its orbit. Each of these crossings will, in turn, send a brief shower of large comets towards the inner solar system. This means that all together several brief showers should occur within the entire event. This not only helps explain why paleontologists find extinctions to be gradual, occurring over several million years, but it also tells us why there are several episodes of species loss, or stepwise extinctions, occurring within each of these longer periods of extinction.

Many people thought the gradual rise and fall of extinctions, and the stepwise extinctions, were inconsistent with comet showers. They believed comet showers could only come from the Oort cloud and that these showers would begin quickly then decline over millions of years. But stepwise extinctions do not last this long. They did not realize that you can get brief showers if you scatter comets from a disk rather than from a cloud, and this is just what you would need to produce stepwise extinctions as well as the overall gradual rise and decline of extinctions every 26 million years.

I did this work in the summer of 1987. I did not do the fundamentals. The idea came from Oort. It was refined by Hills,

Everhart, Fernández, and Whitmire. All I did was insert a population of small comets in the scenario. I took their population of large comets just beyond Neptune and extended it with a large belt of small comets. I simply assumed that the size of the comets decreases from the inside of the disk outwards. So the smaller bodies are found toward the outer rim of the disc, and the larger ones on the inside. And I deduced that there has to be a Dark Planet out there that scatters the comets toward the known planets. Without this Dark Planet, there would be no small comets here today and no storm of large comets every 26 million years.

I am convinced that comet showers are the answer to many scientific mysteries. With a storage disk of comets and a Dark Planet I can produce comet showers and with that I can account for the death of the dinosaurs, the 26-million-year cycle of mass extinctions, the shorter periods of life loss during each of these cycles, and the present shower of small comets into the Earth's atmosphere. These findings, of course, irritated everybody.

I wrote a paper on the topic called "A Hypothesis Concerning the Inner Oort Disc and Its Relationship to Comet Showers and Extinction of Species" and sent it to the journal *Icarus* in November of 1987. It was not well received. One reviewer commented he could do better sitting at a bar drinking his beer. I was too candid and transparent. That was the problem with my paper. I wanted to get the point across, which so many had missed, that a single undiscovered planet could produce brief comet showers as well as long ones. I should have known better. A paper in which everyone can understand the mathematics looks trivial and gets rejected. This one did.

My first public presentation of the scenario did not fare any better. It was at a meeting of the Division of Planetary Sciences of the American Astronomical Association in Pasadena, California. Many people in the audience responded with ridicule and disbelief. I was being disrespectful of current wisdom in astronomy. Another part of the problem concerned language. I am a physicist. Most of the people there were astronomers. Most laymen would not think so, but there is something of a language barrier between us. I spoke the wrong language. Fortunately, I could still understand the signs on the restroom doors.

And so it was that my efforts to find a source for the small comets led to a mechanism that also explained how a storm of large comets may have been responsible for the death of the dinosaurs. Beyond that, I am sure the most frightening aspect of this new hypothesis escaped no one. If these periods of extinction occur like clockwork every 26 million years, they will, no doubt, happen again. A storm involving hundreds of large comets will once more hit the Earth and devastate life upon it. The good news is that we are presently between two such periods of extinction. It will be another 12 million years or so before another storm of large comets precipitates yet another major wave of extinctions here on Earth.

Chapter 22

The Debate Comes to an End

Sometimes while working in the early hours of the morning I amused myself by thinking of the nature of this seemingly deadlocked struggle over the small comets. During my more cynical moods, I thought of several brontosauri that made the effort to enter a quicksand swamp to overwhelm a comrade who was foolish enough to wander there in the first place. Their comrade's name might have been Frank, and those entering the swamp could have been Donahue, Hanson, and the others. One named Dessler could have pointed the way to the conflict. Along the edge of the swamp would stand a multitude of other creatures with lesser capabilities and courage but all secure in the final outcome since the struggle was so one-sided. They would throw a few rocks at the hapless comrade without knowing why. The brontosauri would thrash and struggle and finally sink slowly into the swamp. So it was with the small comet debate that took place in *Geophysical Research Letters*.

Alex Dessler was caught in this debate in a very peculiar way. He was the editor of the journal at the center of the conflict and the small comets had made him famous. He decided to publish our two papers against the advice of his referees and sometimes people held him more responsible than they held me for all the fuss the small comets had created. He probably thought that the debate on the small comets would end quickly, one way or another. But it lasted well over a year.

Dessler received so much heat for publishing our papers that at the end of his first year as editor he felt obliged to explain why he was publishing controversial material that sought to overturn conventional scientific wisdom. "The importance for scientific progress of the occasional new idea that proves correct is out of all proportion to their number," Dessler wrote in an editorial. "Everyone would agree that not all controversial ideas are wrong. Because it is not possible to tell in advance which new idea is correct, it is best to get new ideas into the open literature where they can be discussed, attacked, tested, or supported as the will of the community and the soundness of the idea dictate. This is the way science has advanced, and no better way has been demonstrated."

Dessler ran this editorial at a time when he thought I was at least partially wrong about the small comets. On the one hand, he believed that the atmospheric holes in our images were nothing more than instrumental noise, yet, incredibly, he thought that there could still be such things as small comets. No doubt Donahue's paper on cometesimals had, at one point, helped convince him of that.

"Scientists would like to think that they are steely-eyed, going after a goal in a nice straight line," Dessler said at the time. "Forget it. There's serendipity and good fortune involved, too. A good man makes his accidents turn out well. And here, I think, is a good example. I think Frank is going to turn out to be not quite right. Nevertheless, it will have done a lot of good for science, advancing it in a left-handed way." At the same time he wrote to me saying that if he lacked objectivity, his beliefs were skewed in my favor. "Only if the small-comet hypothesis is verified will my editorial decision to proceed with the publication of your original papers be vindicated," he said. "All editors are biased toward vindication." He realized that his future hung on the fate of the small comets.

There has always been some tension in my relationship with Dessler. The crucial episode goes back some twenty-five years. Dessler was just starting the space science department at Rice University. Everyone at the time was trying to duplicate the success of the Iowa program. The first satellite, Explorer I, was worked on at Iowa and so was the first successful space probe sent past the lunar orbit, all under the guidance of James Van Allen. We became a center of excellence in space science and good students were applying. Other universities soon wanted their own space science programs.

To quick-start the program at Rice, Dessler began hiring people out of Iowa. He decided, in fact, to gamble his space science program on a person he had persuaded to come to Rice from Iowa. I felt that Dessler was making a mistake. I told Dessler that. So did Van Allen. But Dessler to this day says we did not tell him that this person, who shall remain nameless because he has suffered enough as it is, might be a problem. This individual suffered a nervous breakdown and within a few years the Rice satellite development program collapsed. The episode was something of an embarrassment to Rice University. I think they expected too much, too quickly. But just a couple of years ago, at a dinner in Arizona, Dessler was still swearing at me. He insisted that the debacle had been our fault.

During the debate on the small comets Dessler and I corresponded and talked on the phone constantly. When he accepted those papers, it was like saying, "I do." We had something akin to a close marriage for three years but not a pleasant one at that. He often gave me advice and I think he meant well by it. There were times very early in the debate when I thought he was really worried about my health.

I began to feel the pressure after the first six months. I was spending hundreds of hours on the Replies. It was my obligation. I had made a terrible mess and I had no choice but to defend my actions. That was the penalty I had to pay. But preparing those Replies very nearly did me in. I would work until the wee hours of the morning and then Sigwarth would come in later and check my work. He would redo all the calculations to see whether I had made any errors. Sometimes I did. Sigwarth is very good.

Towards the end of the debate Dessler had us rewrite the Replies over and over. That did not help matters. He was putting on the pressure because he did not like what we said. He also had people rewrite their Comments. He was trying to gain control over the debate that had long slipped from his grasp. This was particularly true with the last published Comment, the one by Wasson and Kyte concerning the composition of the small comets.

I had a little trouble with Wasson and Kyte at the beginning because they made fun of small comets. They likened them to the "living comets" proposed by the well-known British astronomer Fred Hoyle and suggested that the small comets might be coated with live viruses. Hoyle and his colleague, the Sri Lankan astrophysicist Chandra Wickramasinghe, have proposed that living cells shed by comets marked the beginnings of life on Earth and that comet-borne viruses and bacteria continue to arrive here from space and may be the cause of sudden, widespread epidemics on Earth. Most scientists have not taken these ideas seriously. But I think that Wasson and Kyte later realized that perhaps their analogy was not so funny and removed the remark from their final draft.

My troubles with Dessler were not so easily resolved. He wrote to me complaining so much about the quality of my Reply that I wondered if he were jockeying for a spot as co-author with Wasson and Kyte. But his real objection, apparently, had to do with the length of the Reply itself. He pointed out what he thought would be "a nice, appropriate response," crossing out all but the first five sentences of my entire four-page Reply! I was flabbergasted. Dessler also seemed to be co-authoring the Reply.

So I rewrote my Reply and kept it down to the required two pages. But Dessler was still not happy. He seemed fixated on associating the effects of small comet impacts with nuclear explosions, accused me of making changes in my model, and he, too, began comparing me to Hoyle. "As I see it," he wrote, "you are painting yourself into a corner, much as Fred Hoyle has. You ought to think about the parallel between you and Hoyle. If the comets exist, you will get full credit if you answer Wasson and Kyte simply by saying that your comets are different from their comets in that your comets have virtually no dust and they assume 50 percent dust content. If the small comets exist, it doesn't matter what you say, you are going to win a Nobel Prize."

I again shortened my Reply, told Dessler why small comet impacts were not like nuclear explosions, and explained that my model did not change. But I did not book my ticket to Stockholm. Nobel Prizes are won for true advances on the frontiers of science and not for generating turmoil in our perceptions of Earth and the solar system, as I have done. I consider such talk about the Nobel Prize nonsense.

I went on to tell Dessler that there was only one similarity between Hoyle and myself. The phenomenon we each proposed profoundly affects the Earth in the present. The reception that such proposals receive among scientists is significantly more intense than those that concern some event in the remote past or the far future. They threaten our beliefs that Earth is hermetically sealed, isolated from the rest of the cosmos. Otherwise his comparison to Hoyle was like comparing apples to oranges. My attempt to estimate properties of the small comets was clearly not the same as "painting myself into a corner."

It became clear to me that Dessler was not familiar with work being done in many of the important topics that were being brought up in the course of the debate. I do not think that Dessler ever understood the small comets. He was convinced that the comets had to stop at high altitudes and leave too much water up there. He was certain that the break-up of the comets at high altitudes would form ice clouds in sunlight that would be easily visible to observers at night. And he was deeply obsessed with the subject of lunar impacts. He was convinced that each small comet collision with the Moon carried the force of a small nuclear bomb. If the small comets were real, then according to Dessler we would see mushroom clouds on the Moon.

But Dessler was wrong on all of these points. I think he was simply unwilling to spend the time doing the reading and calculations that would elevate his conclusions to something beyond personal beliefs. And his apparent fixation with nuclear explosions was understandable as well. I finally went to the library to see how old he was in 1945. It turns out he was a very impressionable seventeen.

Several times during the course of the debate Dessler called and said: "Did you change the numbers? Did you change the size?" But I never changed anything. I made no adjustments. People said that was cruel. Some people made adjustments to allow ten times more comets to fall into Earth than they previously did. Dessler hoped things would change so he could say, "Well, that's the end of it, a mistake was made and Frank is making adjustments." But all along the number and size of the small comets have been the same, their movement in the solar system has been the same, and their composition has been the same. That is all I ever stated about them. That is all there is.

Most people did not work through the series of Comments and Replies that constituted much of the "official" debate on the small comets issue. Most people were only interested in knowing who won. It was a natural curiosity. But no one won or lost. My critics did not win because they did not have any solid facts to show that I was wrong. On the other hand, I did not win either. I did not have any piece of evidence to show that I was absolutely right. What I had done, however, was present evidence which, when taken all together, favored the existence of these small comets.

It was a long, tough year and a half. Dessler knew what he had put me through. At the end of the debate he sent me the "Alexander J. Dessler Award" in recognition of my "truly amazing array of responses to criticisms of the small-comet hypothesis." But the humor only hid the bitterness that would soon surface.

In November of 1988, Rice University had a party to celebrate the twenty-fifth anniversary of their space science department and Dessler's sixtieth birthday. Several people were asked to make contributions. One person ended up giving Dessler a one-way ticket to Venus for his editorial role in a vicious debate about whether or not there is lightning on Venus.

I, too, was asked to contribute and looked for something amusing yet pointed to send Dessler. So I had my secretary, Barbara Hermeier, pick up a copy of Clement C. Moore's "A Visit from St. Nicholas" in the children's section of the Iowa City Public Library. I then sat down with a can of Diet Pepsi and reworded the poem. My satire of Dessler's misunderstandings on the small comets issue was read at the party.

'Twas the night before Christmas and all through the Texan bog
Not a creature was stirring, except Alex and his dog.

His dog murmured something about small comets falling this night
And Alex responded with hopes that this could not be right.

Alex was nestled all snug in his bed
While visions of giant snowballs danced in his head.

And with visions of mushroom clouds hovering the moon
Alex settled down into a feverish swoon.

When out on the lawn there arose such a boom
Alex sprang from his bed and out of his room.

Away to the door he flew like a flash
And out into his yard he made with a dash.

The moon on the giant pile of new-fallen snow
Gave a luster of midday to objects below.

When, what to his wondering eyes, should happen,
As the surface of the heap of snow seemed quickly to blacken.

With blinding flashes in the air so lively and quick
Alex knew in a moment that he must be sick.

More rapid than eagles the small comets came
And Alex eagerly offered each comet a name.

Now, Soter! Now, Donahue! Now, Rubincam and Hanson!
On, Davis! On, Nakamura! On, Chubb and Wasson!

To the pile of snow Alex quickly went
To see how large a crater the small comet had sent.

Now, dash away! Dash away! Dash away, all!
There was no crater there even so small.

So up into the sky Alex anxiously peered
To see from where these small comets were steered.

The legions of small comets at large distances could not be found
And Alex responded with comments not so profound.

The red point of Mars grinned down from the sky
Why aren't you blue, Alex said with a sigh.

His clothes were all tarnished with ashes and soot
As he kicked the comet debris with his foot.

With a wink of his eye and a twist of his head
Alex dashed into his house to get into bed.

And laying his finger aside of his nose, with a grin on his face
He dreamt of all the current wisdom that he would erase.

*   *   *

The small comets story would outlast Dessler. His term as editor came to an end on the last day of 1988. He applied for another term, but did not get it. He probably thinks the small comets controversy did him in.

The small comets are now all that Dessler has left. He has been giving talks on the small comet hypothesis. The same so-called snags in the hypothesis still trouble him. But he no longer discusses these issues with me directly. For a while he sent me letters, signed "Tex." Tex is Dessler's dog. "I'm still convinced that the small comets exist, but I am having trouble convincing my master," Tex wrote early in April, 1989. Tex obviously represents that part of Dessler that still believes in the small comets. To convince his master, Tex wanted "any good physical arguments supported by calculations." Dessler had enclosed a photo of Tex looking up in the sky, ostensibly for small comets.

"Now I know what poor Galileo went through, and you too, poor guy," Tex added. "What should I do? I have considered the unthinkable (i.e., biting him), but I am his best friend, so that is out of the question." I found Tex to be a very perceptive dog and thought his master should reward him.

Two weeks later there was another letter. It began "Dear Mr. Excitement." There was "a disturbing development," Tex wrote. "We have another dog—just a six month old puppy—a female named Spot. She is a real nothing of a dog." But the real problem, Tex explained, is that Spot "has been convinced by my master that small comets do not exist."

I did not reply.

Chapter 23

Astronomers Enter the Fray

To prove to anyone that the small comets really did exist still required the final test. The small comets would have to be seen with a telescope. It was the only decisive way to resolve the issue. It was the most important thing anyone could do. We had to look with a telescope and get a picture of one. This search would constitute the next and final phase in the discovery of the small comets.

Steven Soter planted the seed for this search back in February of 1987. It was not his intention, I am sure, but it was thanks to Soter that someone eventually went out there and looked for the small comets with a telescope. At the very least, he made it happen a little sooner, he gave the idea a little fertilizer.

Soter was an astronomer with the Center for Radiophysics and Space Research at Cornell University in Ithaca, New York. Thomas Gold, the noted astronomer, was also there. Gold, you will recall, had been very upset by my small comets proposal. Perhaps he felt I was treading across his personal territory. So it did not surprise me to see Gold acknowledged at the end of Soter's Comment that ran in the February 1987 issue of *Geophysical Research Letters*. But Soter did bring up a good point: We now had the technology, specifically new sensing equipment for telescopes, capable of spotting these objects.

Soter believed that many small comets should have been observed by telescopes used to search for man-made satellites. One such system is an observatory southeast of Socorro, New Mexico, which served as the prototype for the U.S. Air Force's

Ground-Based Electro-Optical Deep Space Surveillance (GEODSS) network. The observatory consists of two telescopes mounted side-by-side and equipped with a special television camera that permits high sensitivity readings onto videotape. This sensitive television camera, unlike photographic film, allows a telescope to spot rapidly moving objects at low-light levels.

The Air Force has now deployed four such state-of-the-art systems around the world and uses them primarily to track foreign satellites. The Space Surveillance Group of MIT's Lincoln Laboratory developed the system for the Air Force during the late '70s and early '80s. Laurence Taft, an astronomer, was in charge of testing the first of these telescopes in New Mexico. He made a lot of measurements of artificial satellites, satellite debris, and meteors, but he never conducted any organized search program. All he had were notes of his observations, such as "I saw an object." Taft himself admits that "precise records were not always kept."

Taft spelled out the capabilities of the telescope, as well as his results, in a handful of papers. In one test, Taft made a count of man-made satellites and their debris. In the telescope's "deep space" search mode, at altitudes above some 900 miles in other words, Taft spotted about one piece of debris, as small as eight inches, per hour. In a search for near-Earth debris, at altitudes below about 900 miles, conducted in January and February of 1984, Taft found eleven times more objects out there than were cataloged by the North American Defense Command. These objects were as small as four-tenths of an inch.

Looking back over Taft's results, Soter came to the conclusion that if the small comets did exist, then "this system should have detected hundreds of them, even if they are as dark as the least reflecting known objects in the solar system." In the deep space mode, Soter said, the observatory, operating at its limit of detection, should have spotted these objects as far away as 62,500 miles from the Earth. He calculated that, at best, the system should have been able to pick up some 250 small comets per hour or, at worst, about forty per hour.

But when Taft used the telescope in this mode for more than 200 hours during 1977 to 1981, mostly in September through October and in February through March, the detection rate was only about one object per hour. And all of these objects, said Soter, could be identified as belonging to distant man-made satellites, not comets. He thought that even if the comets were very small and reflected very little light, the system should still have picked up several per hour. So, Soter concluded, it "seems unlikely" that natural objects as large, close, and numerous as the small comets, could have been overlooked.

But Taft's actual work left many questions unanswered. If you look at his papers, it seems he saw objects he could not really identify. These objects had no light curves. A satellite turning and reflecting back sunlight would appear to blink on and off—that is known as a light curve. Objects without a light curve are either not spinning or not artificial. In any case, Taft never identifies these objects. Many of the objects he saw at large distances from the Earth did not exhibit the periodic variations in brightness of a rotating spacecraft. He simply assumed they were debris from man-made satellites.

I had questions about his work so I called Taft myself. I asked him how many of these objects he had seen at large distances and could not identify. He said several per hour. I had him repeat it. He said three to five per hour. I think he felt awkward. Later, when I began citing his figures, Taft got mobbed by scientists and the press. He told other people he did not see any small comets. To one reporter, he said: "Frank is wrong. There aren't any small comets out there. If they were there we would see them. We have never seen anything that even remotely resembles them in hundreds of hours of observation. So if they are there, then they are really tough to find."

I think Taft felt the pressure. Suddenly he was right at the center of a very emotionally charged debate. It was an awkward position to be in. Anyway, life is always more pleasant in a cheering crowd than as the lonely supporter of an unpopular renegade.

Though Soter's paper got people thinking about telescopic observations of the small comets, his arithmetic was incorrect and so were his conclusions. He greatly overestimated the detection rate of small comets by Taft's telescope. His rates for the detection of the small comets were a hundred to a thousand times too high. He failed to understand the importance of their dark mantles and their motions past the Earth. But even more important he did not realize that some of Taft's observations took place during a period when the number of small comets was at its lowest.

I had known since the middle of 1986 that there were variations in the numbers of small comets over time. We found that there were ten times as many small comets in early November as there were in mid-January. This overall decline also showed variations on shorter time scales, such as a decrease on about December 14th. No one was aware that there were large time variations in the occurrence of atmospheric holes in the Dynamics Explorer images. No one knew because my paper on the subject, which I had submitted to *Geophysical Research Letters* in the summer of 1986, had been rejected by Dessler.

These time variations quite clearly were a crucial element in any discussion regarding the possibilities for detection of the small comets by a telescope. I realized that this was a splendid opportunity to publish a large portion of my previously rejected work on the topic. This upset Dessler tremendously. He kept wanting to restrict my answer to the single point Soter had raised. He insisted that eighty percent of my Reply had to be a direct response to Soter and that only twenty percent could be new material on time variations. Nor would he let my Reply suggest a new research program. It took Dessler a while to realize that the time variations, when tied into detection rates, were very relevant to my Reply. And so it was that my paper on time variations came to be published after all.

I redid Soter's work taking these time variations into account. It turns out that the rate of detection by Taft's telescope for small comets far from the Earth could be as low as one every ten hours to as many as one per hour. These rates are not inconsistent with the rates reported by Taft. His papers even stated that he saw objects with an increasing frequency as he looked out farther from the Earth. And they did not have the light curves that are typical of spinning man-made satellites. The brightness of a small comet will not vary as it spins because its surface is even; it does not have the solar panels and flat metal surfaces characteristic of artificial satellites.

I think Taft spotted the small comets. He just assumed these objects were spacecraft and spacecraft debris. He said these must be old, retired spacecraft that were not spinning because they were not being used. But old or not, once a spacecraft is spinning it will remain spinning for the rest of its life.

I also had to determine the rates for the detection of small comets when the telescope looked for near-Earth objects. The small comets fragment at altitudes somewhere between about 600 and 1900 miles. For at least a brief moment after break-up, these small comets probably brighten. Their speeds are large enough to produce a streak during an exposure time of one-thirtieth of a second on Taft's telescope. My estimate of the detection rates of small comets at low altitudes with this telescope range from about one event every forty hours at 1,900 miles to one event every 360 hours at 600 miles. According to Taft, the hourly rates for streaks due to satellites and meteors in the telescope frames are about twenty events per hour. That meant that one out of every 800 to 7,200 of these streaks would be due to the passage of a small comet. But these streaks were like needles in a haystack and to identify which ones belonged to a small comet would have required a tedious examination of the image frames. Taft had done no such thing.

The work that Soter put me through was not a waste of time. It would be necessary for anyone who later decided to go out and look for these objects with a telescope. Soter's numbers were not right but they were close enough to establish that it might well be possible to detect the small comets with a specially-equipped telescope. They would be difficult but not impossible to find. The search would have to be carefully done. The observer would have to be dedicated to this task. You could not do it as a by-product of some other random search. You had to have a detailed plan.

This public confrontation between Soter and myself also made it clear that even with this new technology, finding these objects was not guaranteed. You could easily waste one or more years of your life doing such a search. You had a chance of spotting the small comets only if it was done the right way.

Within a year the search would be done the right way.

Chapter 24

Where Are You Now, Galileo?

Astronomers never really bothered to search for the small comets,‘though it was, quite clearly, their responsibility. Even more troublesome, astronomers, by their denials, actually hampered this effort. They just did not care to look.

I should have expected it. Astronomers have traditionally shown little interest in near-Earth objects. They have always sought to see as far out into the universe as they could, not as close to Earth as possible. The only searches in the vicinity of the Earth are for satellites and rocket debris. Telescopes are simply not available for any other kind of search. But even if they were available there is a psychological block. There is a presumption that the Earth floats in a void. Other than a few flecks of dust that cause the bright meteors we see when we look up at the sky, the Earth, everyone believes, is isolated.

For a long time even the dust and rocks were hard to accept. For years such eminent bodies as the Royal Society in Britain were skeptical of stones that rural folks claimed fell from the heavens. Scholars were appalled at the ignorance and credulity of people who reported such things. It was not until 1803, with the spectacular meteorite shower that fell at L'Aigle, seventy miles east of Paris, that such a notion actually became scientifically respectable. But even afterwards there were doubters. Thomas Jefferson seems to have been among them. He supposedly said in 1807, "I could more easily believe that two Yankee professors would lie than that stones would fall from heaven."

I was raised believing that the Earth sits here in relative

isolation. There would be a little dust floating around, certainly, and there was always the fear that a large rock might come by and hit us every ten million years or so. But, otherwise, we are isolated. The universe stopped revolving around the Earth long ago. There is nothing coming in from out there. There is no need to look. It did not seem to matter to astronomers that I, a physicist, had evidence of a stream of objects arriving here from the outer rim of the solar system. And never mind the profound effects such a situation would have on our existence, our history, our oceans and our atmosphere, and the outer planets.

Though the will was lacking, the technology to conduct the search was not. There was a new technology available, as Steven Soter had pointed out, that represented a remarkable leap beyond the photographic film used by most telescopes to record their observations. Film is very sensitive but very slow. If a telescope remains pointed at a star for a long time, a half-hour or an hour, the photographic plate collects the faint light from the star and produces a very good picture. You can afford to spend that much time with a star because stars do not streak through the sky. You can sit there and integrate, or gather, the light from the star for an extended period of time to get your picture. But a rapidly moving object passes over the photographic plate in seconds, so the integration time is short, and unless the object is extremely bright, it will not show up in your picture.

My comets, being so small and dark, are only capable of being seen by reflected sunlight. Such faint objects are difficult to find amid the many bright points of light in the cosmos. To find them required a relatively new technology called array detectors, which are attached to the eyepiece of the telescope. Array detectors are made up of thousands of little photoelectric eyes. Being much faster than photographic film, they are able to record fast objects streaking through the telescope's field of view. These array detectors, which are also called CCDs, or Charged Coupled Devices, have revolutionized astronomy.

Still, few telescopes in the country are presently equipped with this rather expensive technology and, among those that are, not all are suited to conduct a search for near-Earth objects. In 1987, one of the few facilities that could seek the small comets was the Spacewatch Telescope at Kitt Peak in Arizona. The telescope is in the hands of Tom Gehrels, an astronomer at the University of Arizona. Gehrels had been drawn into the small comet debate by Dessler, who had asked Gehrels to review a paper I had submitted to *Geophysical Research Letters* in the summer of 1986. In his review Gehrels said that he could not take the small comets seriously because their numbers were a million times greater than what he thought was reasonable.

Gehrels is interested in the minor bodies of the solar system, especially those, such as asteroids, that cross the Earth's orbit. These are potentially catastrophic if the Earth is in the right place (the wrong place!) to get hit as they come by. This is the kind of object that Luis Alvarez, Gehrels, and others believe caused the death of the dinosaurs. They are interested in these objects because they think that one of them once hit the Earth with catastrophic results and might do so again someday.

There is something of a three-way race to discover asteroids approaching the Earth. That is what Tom Gehrels does. That is what Laurence Taft does. That is what Eugene Shoemaker, an astrogeologist at the United States Geological Survey in Arizona, does. Shoemaker and his associate Eleanor Helin, at the Jet Propulsion Laboratory, use the Schmidt telescope on Mt. Palomar. Their telescope has the largest view of the sky. They use photographic plates, which are perfectly acceptable for asteroids because these objects do not move very quickly against the field of stars. With the Schmidt telescope, Shoemaker and Helin discover more asteroids than anyone else.

Steven Soter was convinced that the small comets should have been detected by people such as Gehrels, Taft, and Shoemaker in the searches for Earth-approaching asteroids. But this is unlikely. These searches are carried out by taking at least two pictures sufficiently separated in time to detect the slow motion of a distant asteroid. The apparent motion of the small comets, on the other hand, is sufficiently large (because they must be much closer to the Earth in order to be seen) that these objects are beyond view for at least one of the pictures. So the small comets would most likely not be seen during searches for asteroids, even if photographic film were sensitive enough.

I worried that no one was going to look. Taft had found nothing using the MIT prototype telescope in New Mexico and was out of the picture. Shoemaker could not look, as he lacked the proper equipment, though he insisted that the small comets were not there. That left only Tom Gehrels and his Spacewatch Telescope. Dessler suggested to Gehrels back in 1986 that he look for these objects with his telescope. But Gehrels would not do so. He was certain that the small comets could not be there. He knew the answer without looking.

Gehrels' posture has been the same all along. In a paper he wrote for *Physics Today* in 1985 he prepared a table of the number of near-Earth asteroids and comets. He listed objects weighing from about a hundred thousand tons to a hundred billion tons and their numbers in the vicinity of the Earth. He then extended his table to small-sized objects, for which he had no observations, to find the highest possible number of hundred-ton objects hitting the Earth. He concluded that one would impact the Earth every 100 years. Our estimate was one every three seconds.

Like Dessler, Gehrels also argued that the impact of each of the small comets on Earth would be the equivalent "of several times the energy expended at Hiroshima in August 1945." He could not believe that twenty Hiroshimas per minute were going off some fifty miles above our heads. Of course, he was talking about rocks, not comets. On the basis of what Gehrels knew, objects of this size fell on Earth so infrequently that it was just pointless to use his telescope to look for them. He just did not want to do it. But it is actually quite possible that Gehrels, like Taft, had already detected the small comets during his time on the telescope. He had just not identified them or had assumed they were satellites. The observation of small comets with a telescope would be an infrequent event.

So the rumor spread quickly. It had started out as "Gehrels is not going to look because he knows the small comets are not there." But within two months Gehrels' stance had been transformed into reality. It was rapidly understood by astronomers that Gehrels had looked for the small comets but had not found them. So naturally there was a lack of interest by astronomers. Gehrels, the world expert with one of the few telescopes properly equipped to spot the small comets, had looked and could not find them. Besides, who would believe that astronomers could miss ten million house-sized objects falling into our atmosphere every year, when they can look at Jupiter and newborn stars, and see to the very edge of the universe some 15 billion light years away? They could not possibly miss so many objects just beyond the Earth's atmosphere.

Astronomers should have learned their lesson long ago. Galileo, in his day, could not get people to look through a telescope to see the moons of Jupiter because people just knew that Jupiter could not possibly have any moons. Everyone just knew that all heavenly bodies circled the Earth. If Gehrels had been alive back then, I think he might have refused to look. But now, three centuries after Galileo, Gehrels, who as an astronomer should know better, was saying, in effect: I don't want to waste my time using my telescope to look for small comets because they don't fit in with my thoughts on this matter.

Why had astronomers not learned the lesson of their great mentor? The answer is simple. Like the moons of Jupiter in the 17th century, the small comets today are an emotional issue. Everyone chokes on the subject. It does not follow current wisdom. I think that skepticism is healthy in science because it acts as a filter to quickly sort out the few true findings and hypotheses from the great number that prove to be incorrect. But skepticism carried to the point at which further observations are not made to decide the issue and opinion is substituted for fact is clearly a detriment to the progress of science.

Soon after I had published the first two papers in *Geophysical Research Letters* many people, including Dessler and Donald Hunten, told me to relax. People would go out and look, they said. People would do the experiments and use their telescopes to search for the small comets. All I had to do was sit still for a year or two and it would all be straightened out. I was naive enough to believe that this would happen. Long before George Bush made the term popular in his presidential campaign I adopted a "kinder and gentler" period of my own. But two years passed and no one—other than Donahue, Olivero and Mendillo—did anything. No one looked. No one did the critical thing. Of course, there was a problem. Imagine being an astronomer. What if you should find them? How then do you answer the question that everyone had been asking: How could you ever miss these things? You would be embarrassed.

Someone eventually would search for the small comets. But it would not be an astronomer. And a disaster would have to intervene. It was January of 1986. We were temporarily distracted from the small comets. After nearly a decade of effort, the Galileo spacecraft, the most complex and most sophisticated robot spacecraft ever built by man, was being prepared for its launch to Jupiter in May. I was readying the instrument I have on Galileo that will measure the ions and electrons in the vast radiation zones of Jupiter. Then, suddenly, the space shuttle Challenger went down in flames. There was massive confusion in the Galileo project. The spacecraft had been specifically designed to go up on a shuttle flight.

A month later we pretty well knew the impact this disaster would have on the exploration of space. The shuttle would not fly for a long time, and Galileo would not be launched until the fall of 1989. But, now, instead of about two years to get to Jupiter, it would take six. Having missed its best launch opportunity, the spacecraft would now require a gravitational boost provided by a Venus flyby and two such boosts by the Earth in order to reach Jupiter. It was still worth it. As long as it happened within my lifetime, I could wait.

So could Torrence Johnson, the project scientist on the Galileo project for the Jet Propulsion Laboratory (JPL). He could work on Voyager, which was making its magnificent tour of the solar system. But some scientists were stuck in the project with little else to do. There were good scientists like Clayne Yeates who depended on the launch. He was science manager for Galileo and had been since 1977, when we first met. Yeates, in fact, was one of the key people overseeing the science investigations on the Galileo mission, including mine. There was some worry at JPL whether anyone on the project would stick around. Would they wait three more years for launch and then another six years for the arrival at Jupiter, essentially doing nothing with their lives until then? It was a long time not to do any science.

Yeates found himself with little to do after Challenger blew up. I sat in his office right after the shuttle disaster listening to him wonder whether he was going to stay on the Galileo project for all the years it would take to recover from this tragedy. It was questionable. He wanted something interesting to do. Of course, I had just the thing. Yeates was curious about the small comets. He had already seen the Comment by Steven Soter and had become interested in using a telescope to search for them. Yeates is an excellent scientist and is willing to look at new problems. He is able to work with equations and is capable of gaining an insight into their meaning. He is also bright and strong-willed and unlikely to break under pressure.

Casani, the project manager for Galileo, was really interested in having Yeates get into something so they would not lose him. He was too valuable to the Galileo project. So in the summer of 1986 Casani asked me to talk to Yeates and encourage him to work on the telescope problem so that he would not notice all these years going by in his life.

Yeates agreed. But he got more than he bargained for.

Chapter 25

The Great Search Begins

"I thought it was pretty wild." Clayne Yeates's first impression of the small comets was no different than anyone else's. "There were a lot of reasons that made it seem unlikely."

But unlike everyone else, Yeates decided to do something about it. "It looked like a real challenge to make the telescope observations because I could either rule out the small comets with a negative result or I could show that there was a population of these objects. It seemed worthwhile to do and nobody had really ever done it before."

Yeates realized that the one telescope capable of detecting very small, near-Earth bodies was actually geared up and available at the time. This was the Spacewatch Telescope located at Kitt Peak and run by the University of Arizona's Tom Gehrels and Jim Scotti. So in March of 1987, after familiarizing himself with the properties of this telescope and spending many months trying to figure out how best to use it in a search for small comets, Yeates approached Gehrels about such an investigation. Gehrels first responded that these objects could not possibly be real. He had long ago made it clear that he did not wish to spend his time looking for them.

Yeates recalls: "We were sitting in a little refreshment room at the University of Arizona at the time, having a Coke or something, and there were a bunch of people in the room. So I said, 'Hey, Tom, we're scientists aren't we? Maybe we ought to look. We should be objective about this and just take a look. It would

be worthwhile if we could show that these things aren't there.'
Then Gehrels changed his tune and said, 'Well, maybe we ought
to look at this.'"

Some time later, during a telephone conversation, Gehrels
asked Yeates how he planned to fund the search: Can you get
some money? Yeates asked how much. Forty-five thousand
dollars, Gehrels replied. Yeates felt the cost was a bit steep for a
couple of nights of observation but told Gehrels he would secure
the necessary funding.

Yeates then spoke to the influential Donald Hunten at the
University of Arizona about the idea, and together with Hunten
and Gehrels, Yeates wrote a proposal to obtain the money from
the Jet Propulsion Laboratory's (JPL) Discretionary Fund. This
was a kind of cash machine available to the JPL director for
interesting projects that need quick funding. As it turned out, JPL
director Lew Allen and his chief scientist Moustafa Chahine
were both quite interested in the search, so Yeates got the money
to rent Gehrels' telescope.

Meanwhile, Yeates completed the details on his scheme to
detect the comets with the telescope. "A lot of astronomers were
making offhand comments like how easy it would be to see these
objects if they were there," says Yeates. "But the more they said
how easy it was, the more suspicious I got. So I started to do a
bunch of calculations to see just how easy this was to do. And
one of the first things that became clear to me was that this was
not very easy at all. It was very difficult in fact."

Yeates knew there was only one way to find these small, very
dark, fast-moving objects. He had to bank on our predictions
from the atmospheric holes, concerning their number, their
speed, and their direction of motion past the Earth. Otherwise
there were just too many possible ways to aim the telescope. But
to get a snapshot of these objects in the Earth's vicinity, Yeates
concluded that the telescope would have to be operated in a
novel fashion. He would have to move the telescope at a rate that
matched the predicted motion of these objects as they passed by
the Earth. He would have to aim the telescope as he would a rifle
in shooting skeet. Only in this case the clay targets were small
comets.

This was the only way, he realized, to hold the objects in view long enough to see them. The objects would be very dim, but by skeet shooting with the telescope, the small comets would be tracked long enough for the telescope's array of photoelectric eyes to have a chance to accumulate enough light from these objects so they could be seen. But without the basic information on the objects' speed and direction, Yeates would have been hard pressed to find these comets. If you are at a skeet shooting range and you have no idea where the clay targets are coming from, or how fast, chances are you are going to miss them.

Yeates counted on our predictions and did all the calculations and worked very hard at it. He did it all himself. I did talk to him from time to time and I checked his calculations. But this was his work. He was the one who had the perseverance to look up all the numbers, find all the necessary information, design the aiming of the telescope, and conduct the actual search. He spent a year doing this. And everyone thought it was a trivial thing to do.

But when Yeates explained to Gehrels how he planned to use the telescope to search for the small, Earth-approaching comets, Gehrels was aghast. He insisted that this was not the way to look for these objects with his telescope. Gehrels wanted to scan the sky for them just as he did for asteroids. He would aim the telescope in a fixed direction to catch the objects as they headed toward the Earth, down the length of the telescope, so to speak. Yeates explained why this scheme would not work. The chances of spotting a small comet this way were almost nothing. Yeates thought they were far more likely to spot these objects by moving the telescope across the sky in such a way as to catch the small comets as they sped by the Earth at a distance almost halfway to the Moon. Gehrels did not seem to understand what Yeates was trying to do and a heated discussion ensued. In the end, Yeates refused to fund any search using a scheme other than his own. Gehrels conceded.

Meanwhile, Yeates submitted a paper to *Geophysical Research Letters* on the method he would use to search for the small comets. Gehrels' 36-inch Newtonian telescope would be pointed away from the Sun, eastward near the horizon at sunset and westward near the horizon at dawn. And the telescope would have to be used at its maximum range, or threshold, which is about 85,000 miles, a distance almost halfway to the Moon. This would allow the observation of the largest possible area of the sky and increase the likelihood of spotting the small comets. The shutter on the camera would remain open for twelve seconds during which time the telescope drive would be shut off. This drive normally allows a telescope to follow the motion of the stars across the night sky. But by shutting off this drive, the Earth's rotation would provide just the right motion to "skeet shoot" with the telescope.

Yeates' skills at this kind of cosmic skeet shoot were unknown but if he should chance to almost lock onto a passing small comet, the object should show up as a dim, short streak in his images. It would only be as bright as a star of the 18th magnitude, which is about one hundred thousand times dimmer than what the naked eye can see in the sky at night. But by using the telescope in this way he thought he should spot as many as one or two small comets an hour.

The reviewers did not like Yeates's paper. One referee criticized the fact that "very restrictive assumptions are made by the author" regarding the characteristics of these objects. What the referee failed to understand is that this search was conceived as a test of the small comet proposal. If you are going to search for objects it is good to know what they look like and where to find them. If the small comets existed, their character and orbits should be as I had described. Any other assumptions regarding these objects would not be a test of my proposal. Incredibly, the referee even objected to Yeates's use of the words "small comets."

The other referees also seemed to misunderstand the paper, including Eugene Shoemaker. He found the detection rates Yeates had calculated too high. "I went through my own calculation of the detection rates with the Spacewatch system...about eight years ago," Shoemaker wrote, "and concluded that this much more powerful system should yield a few objects per year with diameters near [forty feet], assuming about 500 hours of scanning per year." Yeates estimated something on the order of one or two per hour. Shoemaker, like the other referee, seemed ignorant of the fact that the search was a test of my idea, not Shoemaker's.

Shoemaker's review also included the stock answer. The Spacewatch Telescope, he noted, had been operated for a great many hours without finding any near-Earth objects. "The failure to find Frank's microcomets with the present accumulated observing time on the Spacewatch Telescope," he said, "is a pretty good indication that Frank is wrong." There was no need for a search, Shoemaker seemed to say, because he already knew what the outcome of such a search would be.

The referees saw no need to publish a paper on how to look for the small comets with a telescope. Dessler agreed with them and rejected it. But he indicated to Yeates his interest in publishing such a paper along with the results of a search. It was a reasonable decision.

In late October 1987 the search began. But the first attempt never got off the ground. Gehrels and Yeates tried a test run together, but the array detector did not work and nothing was accomplished. The next available date on the calendar was the 20th and 21st of November. Gehrels was in India at the time, so Yeates and Jim Scotti took two full nights of exposures. "It was a pretty crude system," Yeates recalls. "We had to set the telescope in motion by pushing a button with your hand and then push another button by hand that would open the shutter on the camera. It was not all computerized. We had to sit there all night."

Another run was made on January 23, 1988, by Scotti alone this time, though Yeates was in contact with him at the telescope to insure proper conduct of the experiment. Following Yeates's instructions, Scotti also took calibration pictures of stars with known brightness. These would serve as points of comparison that would allow Yeates to estimate the brightness of any small comets that appeared in the images.

The runs had now been completed. The exposures were on computer tape and needed to be analyzed. But Gehrels did not seem interested. Later, Yeates would have to make copies of all the observation tapes; Gehrels wanted to erase his, apparently to save a few dollars with their reuse. The analysis could not be done at Arizona anyway, as Gehrels had no data displays that could be used to look for extremely dim objects. We did it at Iowa. So did the Jet Propulsion Laboratory. But it would have been extremely expensive for Yeates to hire the help necessary to do the analysis at JPL. So I made our facilities at Iowa available to

Yeates, and Sigwarth also offered to help. He was working on his master's thesis at the time and I thought that after Yeates was finished perhaps we could get some of his data and extend the analysis of these images as part of Sigwarth's doctoral thesis.

Yeates came to Iowa a number of times in the winter of 1987-88. He searched the images himself, using the computer programs that Sigwarth had designed to display them on the screen. He spent hours looking at the images, trying to identify every single item in them. It was an exhausting job. Gehrels had known that it would take an immense amount of time to process images of this kind. Yeates did not think he was going to find anything. But he thought the search was worth gambling yet another year of his life.

Yeates studied each picture in painful detail. Consider that each picture is made up of over a hundred thousand pixels or photoelectric eye readings in the array detector. It took a while simply to get used to looking at all the information they contained. The most obvious features were the horizontal streaks of the same length that were due to the motion of stars across the image. These were about a hundred pixels in length. There were also dead spots in the array detector. These were pixels that had failed to operate and were always dark.

Other spots in the array detector were essentially overworked. They were hot and not functioning, but they were bright and unmistakable. There were also bright vertical lines and features in the lower right-hand corner of the images due to other flaws in the array detector. These were always there. And then there were dark patches in the images due to ordinary dust specks on the telescope mirrors and array detector. It took a while for Yeates to learn what and where everything was. It was a matter of endurance. He nearly went blind doing it.

Sigwarth did all he could to help Yeates. Occasionally, I would go up to the image laboratory in the early hours of the morning and see how they were doing. We would look at the images projected up on the screen. After eliminating all the known objects and false signals, sometimes a short, faint, unidentifiable streak remained. The first one I saw was really a thrill. It was actually up there on the screen. The streak had the motion of a small comet and the right brightness. Yeates had only to find out how many of them were out there. Their number had to agree with the number of the atmospheric holes. Of the 1,200 images he took, Yeates analyzed 171 images that way. Thirty-six streaks of unidentified objects appeared in these images.

Yeates considered, then eliminated, anything that could possibly masquerade itself as a small comet. Satellites caused long streaks, much longer than a star. And most often they would be crossing from north to south across the field of view. Besides, the telescope had deliberately been pointed well above the high geostationary orbits, those satellite positions that remain fixed in the sky above you throughout the day and night. And anything in low orbit would cross the sky very quickly, so that also was not a problem.

Earth-orbiting debris from exploded rockets and old satellites were also taken into account. Yeates had been in touch with Don Kessler, an orbital debris specialist at NASA's Johnson Space Center, who had furnished him all the relevant orbits and positions but none had been in view with the telescope at the time of the observations. Nor could the unidentified streaks be accounted for by meteors. If these streaks were meteors, then the meteor rate in the atmosphere would have to be a hundred million times more than that presently known to account for them.

Yeates also identified the effects of cosmic rays, those charged particles that travel at nearly the speed of light and come from remote places in the galaxy. And he had to make certain that the unidentified streaks in the images were not being caused by the array detector itself. In order to be able to recognize such false streaks, Yeates had Scotti do some runs of the array detector without the telescope. He put a black hood over the array detector, oriented it in the same direction as they had the telescope during the actual observational runs, and opened and closed the shutter on it. Cosmic rays would still penetrate the black hood and hit the array detector and any flaws in the detector itself would still show up.

And, indeed, cosmic rays produced large signals in a few pixels and gave a short little streak. These were easily identified in the images. Also unmistakable were the short bright streaks produced by high speed electrons emitted by the slightly radio-active material in the array detector itself. All the standard tests were done, in other words, and then some. In the end Yeates showed that the possibility that he had made an error in finding these objects was less than one in ten million or so.

But Yeates did not limit himself to a mere visual inspection of the images. Once he had identified possible comet streaks by eye, he conducted a rigorous analysis to establish that these streaks were due to the actual presence of an object. Features that repeatedly showed up in each of ten pictures—five taken before the picture in question, five afterwards—were subtracted in order to identify the temporary streak of an unidentified object in the image. In this way the exact positions of all the candidate comet streaks were located. The brightness of all the pixels surrounding each candidate streak were then averaged and compared to the average brightness of all the pixels in the candidate streak itself. The pixels in the streak turned out to be considerably brighter than the background around the streak. The streaks, in other words, were real.

Yeates had come here, looked at the images, and found the streaks from the small comets. He had sorted out everything else. He knew where the blemishes were in the images, he knew where the cosmic rays were, he knew where the stars were, and he knew where the satellites were. If there were no small comets out there, there should have been nothing left on those images. But the streaks were there. "The whole thing worked beautifully," Yeates said.

The streaks left in the pictures by the small comets were usually about ten pixels long. Granted, they were fairly dim streaks, but that is exactly what we had expected. After all, these objects were black balls measuring just tens of feet in diameter and travelling at about 20,000 miles per hour at a distance almost halfway to the Moon. Yeates now had his streaks and could calculate how many of these things were out there. And the mind-boggling thing was that the number he calculated was within a factor of two or so of what we had predicted for the atmospheric holes. His accuracy may have been only a factor of two, or maybe ours was in estimating the number from atmospheric holes. Anyway, what does a factor of two mean when we are talking about a million times more than the accepted number of objects of that size?

Yeates was stunned by what he had found. He had not expected to see streaks. All he had to do was look at a hundred images, find no streaks, and the small comets would have ceased to exist. But he found them. And the objects had the right motion in orbit, the right speed, and they were about the right size and darkness. It was fantastic.

Chapter 26

Double Trouble

Even before the pictures taken with the Spacewatch Telescope had been analyzed, Gehrels had drafted a paper on the small comet search. According to Yeates, the paper concluded that these objects had not been found. There were a few blanks left in the paper for the numbers, such as the number of exposures, which Yeates was supposed to fill in. And the paper listed Gehrels as the first and main author. Yeates was relegated to second author, even though the search and its design had been his idea from the very beginning.

This did not go over well with Yeates. He was very angry. It was one thing to say you were not going to look because you knew the objects were not there. But writing up a negative paper even before you did the analysis and putting yourself as first author, when someone else had done all the work and rented your telescope like a Hertz car, was too much.

Gehrels had sent Yeates a copy of the paper in January of 1988 and asked if he was ready to announce that they had not seen anything. "I was shocked that he would do such a thing without analyzing the data," says Yeates. "You could not just sit at a terminal and look over the images. There was a lot of analysis to be done."

The scientific community had learned of the search for the small comets by Yeates and Gehrels the previous month, at the fall meeting of the American Geophysical Union (AGU). So for months that followed reporters pestered them for the results of their search. "Gehrels was telling the press he didn't see anything," says Yeates. "I think he led people to believe that he had analyzed the data." The press would then call Yeates, who was reluctant to release his findings until his work was complete.

But the pressure finally led to the release, at the end of March 1988, of his preliminary results. A press release issued by the Jet Propulsion Laboratory on the next to the last day of the month said that Clayne Yeates had "apparently confirmed one of the decade's most controversial astronomical theories—that millions of small water-bearing comets strike the Earth's atmosphere every year."

Two months later Yeates went on to present his findings in public, without Gehrels, at the spring meeting of the AGU in Baltimore. Yeates had by now ruled out all other possible explanations for the unidentified streaks in the images. "What I was left with," says Yeates, "was a class of objects that behaved like Frank's. It was the same number of objects. They had low inclination, prograde orbits. And they were about the right size. I am convinced that these are real objects." The objects Yeates found were actually somewhat smaller than what I had proposed for the small comets. "But," Yeates says, "if these objects were actually slightly darker than we assumed they were, then they could be as big as Frank's small comets."

It was clear to many who saw the photographs at the AGU meeting that Yeates had found something. There was something there that had not previously been accounted for, something truly astounding. But some were equally certain that the streaks did not represent real objects and were mere noise—fluctuations in the data that are due to chance. What no one knew at the time is that we had other undeniable evidence to show that the streaks did represent real objects. These pictures, however, had not yet been fully inspected at the time of this meeting. What we had managed to do was take two different pictures of the same object. And we had done it several times. We had consecutive images of the small comets. They showed *the same object in two pictures.*

The idea of consecutive images had first been suggested by Dennis Matson of the Jet Propulsion Laboratory in a conversation that took place in early 1988 with Yeates and Torrence Johnson, also of JPL. As it so happens, all the work Yeates had so carefully done to show that the objects were real could have been avoided by simply getting two consecutive images—one taken right after the other—of the same object as it sped by the Earth. This was the standard of proof in astronomy. Astronomers generally agree that two images of the same object is sufficient proof of its existence.

Yeates told Gehrels that he wanted to take some consecutive images with the telescope and Gehrels said they could do it in April. They were trying to set up a time when Yeates could get down to Arizona, but on the clear moonless night of April 19, 1988, Gehrels, who had some time available on the telescope, decided to make the runs himself. In one run, he operated the telescope the way he thought you should do it to see the objects if they were there, which just happens to be the wrong way. But in deference to Yeates he took two runs the proper way. Unfortunately, there were problems with the shutter, which seemed to stick occasionally, and as a result only forty-eight good pairs of images were obtained. Gehrels sent the tapes to Yeates with a note saying "If you see them here, I'll be convinced."

To obtain these two pictures, the telescope was trained on a particular spot in the sky, the drive was turned off, a twelve-second exposure was taken that was then transferred onto computer tape, and thirty-six seconds later a second image was taken. After each pair of images was recorded the telescope was returned to the original pointing direction to attempt to intercept another small comet streaming past the Earth. This method was designed to insure that if a small comet was in view of the first image then it should be in the second image, unless the object was near the edge of the first image and it moved off the edge of the second one.

Yeates gave us the computer tapes of what we called the double images, as he had agreed to let Sigwarth and me write up the results. These images had to pass a set of very strict requirements to qualify as two different pictures of the same object. The streaks on each pair of consecutive images not only had to be of the same length and brightness, but they also had to be separated by three streak lengths, which represents the object's motion between frames. In addition, the two streaks had to pass through the same straight line with reference to the stars. So the requirements were highly constrained. The two streaks had to pass a total of four tough tests: direction, separation, brightness, and length.

And they did just that. Each second streak looks like the first in the pair and each time it was found in the expected location based on the time delay between exposures. The streaks also appeared at about the same rate as they had in the single frame exposures. Yeates had found one streak in every five or six photographs. This meant that in forty-eight pairs of double images we should find about seven or eight small comets. We found six. And each time a comet appeared in the first exposure, it also appeared in the second one taken thirty-six seconds later. The chances that the pairs of streaks were due to something other than a swarm of objects near the Earth were one in several million for some images and one in several billion for other images. We calculated that in order to get one pair of streaks that was due to noise and not a real object would require the continuous operation, twenty-four hours each day, of the Space-watch Telescope for at least one year for some of the sightings and more than a thousand years for others. But only a few hours had been needed to take pictures of the small comets. Such is the power of the two picture technique.

I wanted to present these findings at the next AGU meeting that was coming up in December, 1988. As usual, I had to write a short summary of the work long before the meeting itself. Gehrels had taken these images, so as a professional courtesy I put his name on the paper. I had written to him about this but he was in India at the time. His secretary apparently failed to forward his mail. When Gehrels returned, he was upset that his name would appear on the summary in the announcement book of the upcoming meeting. Gehrels wrote an angry letter to the AGU. He wanted his name off the paper. He did not know the results, but this did not seem to matter. I told Gehrels that I would take his name off the paper at the meeting. I explained that I had put it there simply as a courtesy. After all, it was his telescope. And he had taken the pictures.

During this time, Yeates wrote up his findings on the small comet single exposures. Once his papers were completed, he sent a copy to Gehrels, along with a set of slides. Gehrels was upset. "If only you could have shown me your pictures before talking to science writers!" he wrote. Gehrels looked at the slides and so had, independently, Jim Scotti, Uwe Fink, Wieslaw Wisniewski, and Bob McMillan, all at the University of Arizona. These were "all experienced CCD observers," according to Gehrels, and together they had concluded that the features Yeates indicated "could be explained by...noise." Gehrels then begged Yeates to "let up" until he had seen the same streak twice on consecutive exposures. "If it were true," he said, "it would be a major discovery, an upset in our understanding of the solar system." But Gehrels had yet to see anything convincing.

Shortly afterwards, in a telephone conversation, Gehrels told Yeates that he did not want to be a co-author on these papers. Yeates agreed happily. After all, Gehrels had been certain that these objects did not exist, had asked Yeates to pay for telescope time, and did not participate in the analysis of the pictures. He still did not seem to completely understand the search technique that had been employed and he confided to Yeates that he really did not understand standard methods of identifying noise, which means he was not really qualified to judge whether the streaks in Yeates' images were real or not. Yeates later suggested that he read an elementary textbook on the subject. I also spoke to Gehrels and offered to come to Arizona and explain the pictures to him in detail, but he refused.

Later I sent Gehrels one of our pairs of images of small comets. These consecutive images dramatically confirmed the methods used by Yeates to identify the events in his earlier set of single exposures. They also showed that Gehrels never really understood how to search for the small comets. Gehrels wasted half of his observation time that evening back in April looking in a direction which Yeates had explained would yield no small comet sightings. But ironically, Gehrels' mistake had provided further support for the work done by Yeates. No comet streaks, either single or consecutive, were expected from Gehrels' choice of pointing the telescope, and, indeed, none were found.

But the double images failed to convince Gehrels. "We have been in this business a long time," he wrote on October 17th, "making similar observations and discerning radioactive decay images." He then asked why I had only sent him one pair of double images. "Is that all you have?" And why had Yeates released the pictures to the public without showing them to him first?

He also denied saying that he did not understand Yeates's work. "What I did say was: If you have a major discovery and if you show pictures at all, they have got to display convincing images." If we were seeing something, Gehrels suggested, perhaps they were particles from Halley's Comet.

Nonetheless, Gehrels hoped that the small comets "were here to stay, as a new population of the solar system. We will then honor you and Yeates for having pushed them." But he could see no evidence for them and he wanted his name taken off the paper for the AGU meeting.

The pair of pictures of the same object I sent Gehrels was not the only one, I replied. There were, in fact, six such pairs out of forty-eight pairs of images. Opinions as to whether or not the objects are real were simply not relevant. Where is your analysis, I asked Gehrels? He had, of course, never done any work on the images. His casual inspection of the photographs was meaningless. I asked him how "radioactive decays" could produce consecutive streaks in the pairs of images? It was simply impossible.

I then went on to inform Gehrels that it was Yeates's right to release the images. The Jet Propulsion Laboratory purchased the use of the Spacewatch Telescope and Gehrels had not participated in the analysis of the images. I also asked how he could possibly propose an alternative explanation for the objects in the photographs, if he did not even believe the objects were real in the first place. And if he had looked at the size and number of objects involved, he would have realized that his explanation regarding Halley particles was absolute nonsense. I promised to remove his name from the paper.

Gehrels also sent a copy of his letter to Yeates. Yeates sent him a stinging reply. "The complete absence of any quantitative analysis of the data on your part," he wrote, "even though you had the images at your facility, plus your insistence that visual inspection of the images is sufficient to determine if the images represent real objects, leads me to conclude that you are not competent to 'check their validity.'" Yeates reminded Gehrels that he too had long ago offered to come to Arizona and discuss his analysis of the pictures with Gehrels personally, but that Gehrels had refused on the grounds that he was too busy. And Yeates objected to Gehrels' statement that he was "pushing anything." Said Yeates: "I am just presenting experimental results." It was a very trying time for Yeates.

Gehrels replied a week later. "Your reasonings are impressive," he told Yeates. "On the other hand, it would be reassuring if only a few of the pairs of images would be brighter than near the noise level. If there is indeed a major new population in the solar system, it needs a strong confirmation. Astronomers need to know about this situation and they will quickly confirm or otherwise." Gehrels then sent me a brief postcard. "Relax," he wrote. "Life is too short for animosities. The future has the answer."

Gehrels later insisted that the following note be posted with my paper at the December AGU meeting. "If there indeed were as many cometesimal events in the noise of our CCD frames as L.A. Frank and C.M. Yeates believe, I would expect a noticeable distribution in their size and distance such that occasionally we should see a convincingly bright event. The images that have been shown to us are unconvincing, and this includes the observations of 19 April 1988 for repeated events. Anyone can look at the pictures released by Yeates and Frank, and decide for oneself. Before believing the discovery of cometesimals, a new type of population in the solar system, one would want to see a few good images. Independent observations are needed."

There were several problems with Gehrels' statement. He was talking about cometesimals. These were Donahue's objects and they had long ago ceased to exist. What Yeates and I claimed to see in the images were small comets. And contrary to his statement, the sizes of the small comets had been determined from the brightness of the streaks in the images. It was an exciting result and showed that the numbers of larger objects were much less than the smaller ones. And with regard to independent observations, this had been discouraged already by the folklore that Gehrels had looked for, but not found, the small comets.

So instead we posted the following: "We express our appreciation to T. Gehrels of the University of Arizona for obtaining the pairs of consecutive images. Dr. Gehrels declined our offer of co-authorship because he did not want to participate in the design of the telescopic search or the analysis of the images. The fluxes of the small comets are larger by a factor of about 100 million than the fluxes he believes are there but has not searched for."

Meanwhile, Yeates submitted a pair of papers on his search for the small comets to Dessler at *Geophysical Research Letters*. One was a paper on how he had conducted the search with the Spacewatch Telescope, the other was the results of his analysis of the single exposures he had obtained. The reviews that came back were amazing. The first referee was clearly against publication. "What is missing from this paper is a scientific mind-set," wrote the referee. "I see many streaks in the figures. So what?" He went on to say that anyone familiar with near-Earth and deep-space artificial satellites could "easily invent explanations" for these streaks. Of course, not enough satellites have been launched to account for the high numbers of streaks in the images. This referee stopped just short of "inventing" a fleet of UFOs to explain the streaks. Yeates was dismayed by the quality of the comments. He agreed that anyone could "invent explanations," but he explained that the streaks were so restricted in their character that no other known population of objects could have produced the observed streaks than the small comets.

The second referee, Eugene Shoemaker, who prefers not to hide behind a cloak of anonymity during the review process, recommended the publication of the paper describing the way to search for the small comets, but felt that the results reported in the second paper did not pass "the normal criteria for acceptance of observations of small bodies." He insisted that the streaks were just noise and went into a discussion of his findings with photographic plates which had nothing to do with the present search. Yeates used Gehrels' CCD, this array of photoelectric eyes, because film was worthless in such a search. Shoemaker then went on to underscore the necessity of "*multiple* images taken in succession," and ended his review with a bold wager: "I'm prepared to give odds that he will not find any verifiable images, although it is possible that he could get lucky." Little did he know that he had already lost his wager.

Some time later Yeates received the comments of a third referee who Dessler said was a "CCD expert." "This is the only person who really reviewed my paper on technical grounds," says Yeates. In his review, the referee had recognized that Yeates had done "a considerable amount of work on both papers" and felt he should be commended for it. But he questioned the performance of the array detector and had "very strong doubts" about Yeates's conclusions. He could not recommend publication, he said, but admitted that "there is one argument that would convince me of the reality of the streaks. If they could be found on successive images." Yeates prepared a six-page response to the referee's comments and sent it on to Dessler.

At the beginning of September Dessler wrote to Yeates, clearly stating that "for your paper to be accepted for publication, the referees must be convinced that you have seen the same object in two consecutive exposures." This was of "absolute importance," he said. He also asked that Yeates delete all his work regarding the reality of the streaks, as the pairs of images would make this "superfluous." And, quite unbelievably, he also wanted Yeates to eliminate most of the discussion of alternative explanations that might account for the streaks. But it was Dessler's final comment that made Yeates laugh: "Until one of these objects is tracked for several days so that an orbit can be determined, your paper fails the primary test of demonstrating that the streaks are caused by the same objects proposed by Frank..."

Even a high school amateur astronomer would know that a small, dim object going past the Earth at 20,000 miles per hour cannot be tracked for several days. Yeates informed Dessler that his demand was "*utterly impossible.*" At best, Yeates wrote, "an object such as the one I'm observing could be viewed for only several hours during one night if it could be located and tracked." He also told Dessler that he would provide a pair of successive exposures that show the same object, but that the contents of his paper had to remain intact: "The results of the flux, size, [orbital] inclinations, and ranges are all required in order to show the correspondence between the observations and

Frank's hypothesis." The double images simply confirmed the single images and showed they were real, he explained to Dessler, but "nothing more about the physical nature or orbital parameters can be learned from the double images which is not already known from the single images."

When the referees saw the double images, they must have been taken aback, for the rules of astronomy were about to change. Yeates returned from his honeymoon at the beginning of December to find a letter from Dessler informing him that his papers had been rejected. Enclosed was a second review by Shoemaker saying that three consecutive images of the same object were needed for him to believe the streaks were not noise. Yeates, quite understandably, was angry. It seems as if astronomers had decided to change the standard rule of confirmation just for us. It was an honor, I suppose. The existence of these objects could be very embarrassing for Shoemaker. Imagine spending all your time looking for objects that are going to hit us in the future and missing all these objects that are hitting us right now.

So rather than having two images of the same object, astronomers now randomly decided that three were necessary. But I think that if Yeates had three they would have wanted four. And if he had four they would have wanted five. This was not science. This was not a review process. He felt like a donkey being led on with carrots. Yeates was very unhappy about the poor treatment he received at the hands of Dessler and *Geophysical Research Letters*. The papers on the search were supposed to be reviewed within a month and either accepted or rejected. But Dessler strung out the process for half a year. Yeates realized then that it was a mistake to have submitted the papers to Dessler.

Yeates had answered every criticism raised by the reviewers. But with the rejection letter from Dessler and the second review by Shoemaker was a review from a fourth referee, which Yeates had never seen, as well as a short note from the third referee, asking why Yeates had ignored all his points. Yeates had prepared a six-page response! So Yeates called Dessler and asked him about this. As near as Yeates can remember, Dessler replied: "Well, it wouldn't have mattered anyway. I think I forgot to send it to him." Dessler had simply ignored Yeates's defense of his work. This kind of treatment can crush an individual. But Yeates held up. There was nothing wrong with his papers. People were simply frightened of the result.

Shortly after Dessler's rejection letter Yeates received a call from John Horgan at *Scientific American*. Horgan had heard the news from Dessler and wanted to get Yeates's response. Yeates was furious. Dessler had done what no editor is supposed to do. He had revealed what the referees had said about Yeates's papers. The story, entitled "A Snowball's Chance," told of the evaporating support for the small comet theory and explained why Dessler had decided to reject the Yeates papers. Three referees, Horgan wrote, thought that the streaks produced in the images were noise and one believed that the streaks showed the passage of orbiting debris. It was a free-for-all. Donald Hunten later told Yeates that he had the distinction of being the only scientist he ever knew whose rejection of a paper was announced in the press.

The story had gone on to say that Yeates was considering protesting Dessler's decision. And I, supposedly, was considering setting up an observatory in Iowa to look for the small comets. This was not quite true. Actually what I was going to do was look for the brief flashes produced by their impact on the Moon. But it did not matter; Dessler had already sold Horgan the Brooklyn Bridge. Just before publication, Horgan had called to ask me if consecutive images of the same small comet were available to show that the search was successful. I said they were and sent them to him Federal Express. But to my dismay, the pictures were returned immediately with the comment that there was no space in the magazine for the pictures. Instead, the article contained more silliness from Dessler suggesting that "these efforts" would allow the small comets to persist in the public imagination—just like the Loch Ness monster.

Not content to ridicule the small comets in public, Dessler also did his best to keep any mention of them out of the scientific literature. Barclay Clemesha, a scientist at the Instituto de Pesquisas Espaciais in São Paulo, Brazil, provided a recent example. Clemesha apparently had a run-in with Dessler over a paper Clemesha had submited to *Geophysical Research Letters* on a phenomenon called the "Sudden Sodium Layer." The SSL, as it is known, is a formation of thin layers of enhanced sodium concentration that sometimes appears in the upper atmosphere and is not well understood. Clemesha thought the small comets, which he called cometesimals, could contain sodium dust particles and be the source of this curious phenomenon. The usual mechanisms cited to explain this sodium layer, said Clemesha in a letter to me, "are highly improbable, and I still believe that something related to cometesimals is the best of a bad lot. Unfortunately, Dessler only accepted the paper on condition that I remove all reference to cometesimals!"

There was a war going on out there. We had to fight to get our papers published. We had to fight to let astronomers know that these objects existed. If Yeates had gotten negative results his papers would have been accepted for publication within forty-eight hours and he would have been proclaimed a hero. That is the way the system works. But these findings were embarrassing. How could astronomers miss these things? So when the comets were seen through the telescope no one believed it. Gehrels was not convinced. Shoemaker was not impressed. Torrence Johnson, an astronomer at JPL, thought they could be real, but insisted that the rate at which these objects passed the Earth had yet to be determined. Others simply said that, regardless of the evidence, the objects could not possibly be there. Astronomers had forgotten the lesson of Galileo.

No one is going to open their arms to you if you go against the mainstream. My own paper on the double images, which I prepared for the December AGU meeting, was relegated to the poster session, where papers are literally posted in a room rather than presented orally. Some reporters came up to me and said, "Well, this must not be very important because it's down here in the poster session." But I made no fuss about it. I knew they would do that. That is the way the game is played. It did not seem to matter that the appearance of these objects in two consecutive pictures was probably one of the most exciting things to happen in the field of astronomy in the past decade.

Chapter 27

The Turning Point

The search was over. The existence of the small comets had been confirmed. But few believed it. We had won after nine innings but the others insisted that the game go on. There would be no closing time. We would just continue to run up the score.

The 23rd of March 1989 was like any other day. Except that the cosmos threw a curve ball at Earth and astronomers struck out once again. An asteroid measuring about half a mile in diameter had sped by the Earth unnoticed at about 46,000 miles an hour. This boulder missed the Earth by a mere half million miles, which is about twice the distance to the Moon. In cosmic terms, this was a very close call.

Had this huge rock hit us, it would have left a mile-deep crater in the ground the size of the District of Columbia. Had it impacted the ocean, waves hundreds of feet high would have swept over coastal areas and caused widespread destruction. And one day it just might. Scientists are convinced that the asteroid would likely return sometime to hit either the Earth, the Moon, or Mars.

The event itself is alarming. But more worrisome is the fact that this huge rock crossed the Earth's orbit undetected. No one knew about it until almost two weeks after it had passed. The object was found by Henry Holt, a retired U.S. Geological Survey scientist now working on a NASA-funded project to detect and track asteroids. Holt discovered the object in a set of photographs which he and Norman Thomas, a retired astronomer from the Lowell Observatory in Flagstaff, had taken on March 31 using the eighteen-inch Schmidt telescope at the Mount Palomar Observatory in California. When the photographs were processed several days later Holt noticed a streak of light against the background of fixed stars in two pictures taken an hour apart. This streak indicated that the asteroid was moving across the Earth's path and not on a collision course. A NASA scientist told the press that an asteroid heading right for the Earth would look just like a small fixed star and would be even more difficult to detect.

This is how little we know: A mountain-sized object narrowly misses the Earth and essentially escapes the attention of astronomers. It was more than a million times more massive than a small comet, but no one saw it approach the Earth. No one knew it had come so close until it had already passed. If astronomers cannot even see a mountain-sized object as it approaches the Earth how could they expect to see the small comets? Whatever gave them the impression that they would be able to detect something the size of a small house, and as dark as a black sheet of paper, in the Earth's vicinity in the first place?

The close call seemed to mark a turning point in the small comets controversy, though the first faint glimmerings of acceptance began to appear several months before. In January of 1989 NASA had awarded me a $30,000 grant to work on small comets. This is not a lot of money. I get millions of dollars to do scientific research, but until I received this grant I did not have a legitimate license to work on small comets. Of course, the amount of money itself was far less important to me than what it signified. This was official recognition for the legitimacy of small comet research. Though my proposal asked for financial support on several investigations, only one of which concerned small comets, all but the small comet funding was rejected.

The new social situation took me by surprise. A few months earlier I learned that my career apparently had not been sacrificed by my pursuit of the small comets. The University of Iowa, in recognition of my work, and perhaps in spite of the worldwide controversy over the small comets, gave me the Carver/James A. Van Allen Chair in Physics. The Carver is an endowed chair and is considered to be the highest ranking academic position at the university. There are only twelve such chairs. The chair is also named after James A. Van Allen, discoverer of the radiation belts around the Earth and my mentor as an undergraduate and graduate student. I could go on doing what I had been doing. And I would do just that.

Yet it was quite clear to me that the scientific community in the United States was still in such a deep emotional state over the small comets that publishing the results of Yeates's recent telescope search was just not possible in this country. So, instead, Yeates and I decided to submit our papers on the search to refereed European journals with excellent reputations. The result was like a breath of fresh air. The comments of the reviewers were helpful and fair. Yeates submitted his paper on the search and the single images of the small comets to *Planetary and Space Science*. The reviews were favorable and the paper was published, after some minor changes, in October of 1989.

Our paper on the consecutive images was submitted to *Astronomy and Astrophysics*. The referees made a few suggestions and wanted a statement in the paper saying that the observations were tentative evidence for the small comets and that confirmation would require a more capable telescope. They also wanted the inclusion of an alternative suggestion that the tracks might be due to a previously unknown population of tiny moons orbiting around the Earth. I agreed to these changes and the paper was published in February of 1990.

We had successfully bypassed Dessler. But he was still up to his old tricks. His three-year term as editor of *Geophysical Research Letters* ended on December 31, 1988, and his application to continue as editor had been turned down. But Dessler spent much of the following year writing a long paper about why the small comets did not exist. It seemed to me not to be a serious statement but a spoof of science written for the entertainment of the reader. Impacts of objects with the Moon were described colorfully in terms of "tinks," "pops," "booms," and "woomps." And as proof that small comets could not exist, Dessler cited magazine and newspaper reports in which Dessler himself had been quoted as saying that Yeates's telescope observations were rejected as noise. Who was he trying to fool?

James Burch followed Dessler at the helm of *Geophysical Research Letters*. Burch is a scientist and vice president of the Southwest Research Institute in San Antonio, Texas. He is known to be objective and mild-mannered. So in June, when the scientists at the University of Calgary decided to withdraw their names from the joint paper we had been writing on the atmospheric holes seen by the Viking spacecraft, we decided to submit the paper to Burch on our own. Sigwarth and I had put a lot of work into it; we had shown that the chance that the Viking holes were not real was about one in ten thousand. This was good enough for us.

Burch sent the paper to two referees, one who had good suggestions for strengthening our arguments and the second who was more interested in venting his frustrations. He amused himself by comparing the small comets to flying pigs. "As credible evidence for the theory that pigs can fly..." this referee wrote, "one would require considerably more than a selection of blurry photographs...purportedly showing examples of the shadows that pigs have cast on the ground as they flew rapidly by overhead." Burch did not allow us to see these comments until the paper was accepted because they had nothing to do with the proper conduct of scientific inquiry. When we finally did see them, we found them quite amusing. In December of 1989 Burch published the paper showing that the atmospheric holes had appeared in the pictures taken by the Viking satellite. It was a very good month.

Much to his credit Burch was not closed-minded on the subject of small comets. He also published the work of Scott Bolton, a young, capable scientist at JPL who helps control various instruments on the Jupiter-bound Galileo spacecraft with radio commands sent from the ground. While researching his doctoral thesis for the University of California at Berkeley, Bolton found an unusual effect in the data he had examined. The wind of positively charged ions that flows continuously from the Sun into interplanetary space appears to slow down along a portion of the Earth's orbit. This happens almost every year during the months of November to January. Bolton claims that this slowdown of the solar wind is not due to the Sun itself, unless there is some unknown yearly activity, but to some undetected material in interplanetary space. The only source of this material that he can identify is the water vapor that leaks from the small comets.

Our search for the small comets was delayed by the astrono-mers. We were delayed because they would not look. That is something I never counted on. I never thought that we would have to take the extra time to go in and be astronomers ourselves. That we did so certainly means that our findings are less believable than if astronomers had done it themselves. But we had no choice. It was a very bad social situation. Now, perhaps, astronomers will be forced to go in and take a look. If not, the big breakthrough could come from their far more numerous kin, the amateur astronomers.

In the summer of 1989 I was invited to give a talk on small comets to a large gathering of amateur astronomers at the Texas Star Party at Prude Ranch near Fort Davis, Texas. I was imme-diately struck by two things when I arrived. I realized that my home in the cornfields of Iowa was not the world's end; this was. I was also very much impressed by the vast array of telescopes on the fields of the ranch. Though I am not an amateur astronomer, I had been asked many times about the possibilities of looking for the small comets with smaller telescopes, but had been too busy to do anything about it. Now, I had done my homework.

Amateur astronomers whose telescopes have mirrors or lenses measuring twelve inches or larger should be able to sight the small comets. But they would have to know where to look and what to look for. During the course of a day there are two times for observation, each about one or two hours long. One ends about forty-five minutes before sunrise; the other begins about forty-five minutes after sunset. The small comets will be seen at a distance about 2,500 to 4,500 miles from the observer, so the telescope should be pointed in such a way that it is looking for them at these distances, just outside of the Earth's shadow. Inside the Earth's shadow the objects are not illuminated by the Sun and are invisible.

What would you see? Every two hours or so a small, quite dim object will slowly move across your view, as long as your field of view is about four times the size of the Moon. The object will move a distance equal to the Moon's diameter every five seconds or so. Of course, if there are ten telescopes viewing different regions along the Earth's shadow then the number of sightings by all ten are one small comet every fifteen or twenty minutes.

One amateur astronomer came up to me after the talk and said: "We've seen objects like that, but we just never counted them. We thought they were satellites." But the number of man-made satellites that move in this way are very small and if the number of events seen by the amateur astronomers is high, as predicted, then they must be due to small comets. There are thousands upon thousands of suitable telescopes in the world and their combined observing power is awesome relative to what professional astronomers can achieve with their larger, though far fewer, telescopes. But best of all, amateur astronomers are not afraid to look.

Neither am I. My next experiment will be to look for evidence of small comet collisions with the Moon. Many people were under the impression that we were setting up an observatory in Iowa to search for the small comets as Yeates had done. *Scientific American* even said so. But anyone with common sense would have realized that the observatory here could not be used for such a purpose. The collecting area of the University of Iowa telescope is simply inadequate. What we planned to do, instead, was look for the impact of these small comets on the Moon.

During the course of building the next generation of cameras to be lofted into space, we purchased an array of photoelectric eyes that is sensitive enough to look for brief flashes on the Moon with the University's own twenty-four-inch telescope. This will allow us to take rapid exposures of the Moon with enough sensitivity to confirm whether the flashes that have been reported on the Moon are there or not, and if so, how often they occur. This has to be done when the Moon is not too bright, when a large portion of its Earthside face is in the dark. When the Moon is nearly full and bright, its light is likely to swamp any flash of light measuring tens of feet on its surface. We do not have the resolution to see tens of feet on the Moon. But by looking at the darkened part of the less-than-full Moon, we should be able to see the impacts of these small comets on its surface, bright flashes lasting just a second or less.

Others are also looking for novel ways to use new technology to search for the small comets. Among them is Peter Banks, a professor of physics in the Department of Electrical Engineering at Stanford University. He showed in the June 1989 issue of

Geophysical Research Letters that the radiant heat from the cometary water clouds should be detectable with modern-day sensors on low-altitude satellites. This heat radiation, also known as infrared radiation, is produced by the collision of the cometary water as it plunges through the very tenuous atmospheric gases at high altitudes. Banks is quite right. Hot water turns out to be quite bright in the infrared, which is the same type of radiation that you feel from an electric stove. Such measurements have probably already been made, but public access is restricted. Missiles and rockets also eject hot water vapor into the atmosphere and the search for such hot water molecules may have already been accomplished in the early stages of Star Wars. If so, vapors from the small comets also may have been detected.

Another exciting experiment was suggested by a question that came up again and again during the debate on the small comets. How often would the space shuttle and other spacecraft run into them? It was a natural question, so I did some calculations. The results were fascinating. At high altitudes, an impact of a spacecraft with a small comet that has not vaporized would be disastrous. But these comets are small and the collision frequency is low, so an average-size spacecraft would only be struck once in every 50,000 years or so. This means that one spacecraft in every thousand will be struck in high Earth orbit every fifty years. Has it happened yet? No one knows. But some spacecraft have been lost and no one knows why.

The situation in low Earth orbit, where, in other words, the space shuttle astronauts fly, is quite different. They will run into one of these things once in every 200 orbits. But what they collide with are the cometary water clouds, not the comets themselves. The collision of a cometary water cloud with a spacecraft is benign. The shuttle would survive such a collision very easily. Its surface temperature might go up a few degrees and there would be a slight drag on it, but these effects would be nearly undetectable. So the astronauts have probably flown through these things and not known it. But there is a way for the astronauts to see these cometary water clouds.

Enter Stephen Mende, a physicist at the Lockheed Palo Alto Research Laboratory in California. Mende has proposed a very simple, low-cost experiment that would allow the astronauts to see the cometary water clouds as they move across the sky. It turns out that these things can become visible in one of the ways that Halley's Comet has been able to be observed. Halley's has an immense hydrogen envelope around it, produced by the break-up of water into its constituents, a hydrogen atom and a hydrogen-oxygen pair, from exposure to ultraviolet light from the Sun. This hydrogen-oxygen pair is known as the hydroxyl and it scatters solar radiation invisible to our eyes in the ultraviolet. Halley's has been photographed in this light.

The small comets should produce the same cometary hydroxyl as they break up above the Earth's atmosphere. If the astronauts had a camera outfitted with a device that allowed them to look at this wavelength, they should be able to see the clouds of water vapor from these comets. This camera would operate like the nightscope of a rifle but for ultraviolet light instead of infrared. The clouds would look like great spheres plunging through the atmosphere. They would be a spectacular sight.

The problem is that the special hand-held camera for ultraviolet light was burned up in the Challenger accident. But Lockheed has built another one and the subject of what could be done with it came up at a meeting with the astronauts in the fall of 1988. There were some two dozen astronauts at the briefing at the Johnson Space Center in Houston and I was there discussing how they might help take better pictures of the auroral lights. Someone then mentioned the small comets and I said there was a way to see them. There was some snickering, at first, but I think they would love to do it.

Mende and I have worked out how such an instrument could be used to search for the small comets. We think that the astronauts will need several hours of observing time and they will have to point the camera in a particular direction to be able to see these large cometary clouds. With luck they might be able to spot one every seven minutes, but it might take as long as forty minutes between observations. And observations will only be possible when the camera and shuttle window are not illuminated by the Sun or by the dayside of the Earth because this light will blind the camera.

Eventually the astronauts are bound to experience a close approach with a cometary water vapor cloud. It is likely to frighten them. It would start as a bright little dot in the star field and expand quickly into quite a large object. Imagine seeing such a thing coming at you at 40,000 miles per hour. You would see it for about a minute. But if you looked away from the camera momentarily to the point in space where the object was hurtling towards you, your eyes would see absolutely nothing.

Chapter 28

Small Comets and the Future

We have learned much in the past nine years. After being prodded by the appearance of mysterious atmospheric holes in the images taken by an Earth-orbiting satellite, we came to believe that these holes were actually large clouds of water vapor produced by small comets as they entered the Earth's atmosphere. Later, these same atmospheric holes were found in images of the Earth taken by another satellite, Viking. Our claim became increasingly plausible when Dynamics Explorer picked up the presence of man-made cometary material sent up by rocket into the atmosphere. And then, finally, in 1988, the existence of the small comets was confirmed when a search using a telescope found these small dark objects in near-Earth space.

Meanwhile, we obtained additional support for the existence of the small comets and their implications for the Earth. Traces of the passage of these objects through the middle atmosphere—large, momentary bursts of water vapor—were found by meteorologists at Pennsylvania State University. And after an analysis of material from Halley's Comet, European scientists found that comets have the same chemical fingerprint as Earth's oceans, giving substance to our conclusion that the ocean waters were deposited on this planet by an influx of small comets over time. It is also likely, according to researchers at Cornell University, that the basic ingredients for the beginnings of life on Earth come from comets. Besides this incredible bounty, the small comets may also be responsible for a host of strange, as yet unidentified phenomena in our present-day atmosphere, ranging from UFOs to falls of ice from our skies.

The small comets have always existed. What we did was isolate a set of data that allowed us to give them a name. These objects had actually been previously identified as Öpik's dust-balls but were thought to be a peculiar type of meteor. Radar has also probably tracked these objects, but they were once again misidentified and thought to be stone or iron meteors. These small dark clumps of water-snow are also likely to have passed within view of some of our telescopes, but astronomers, it seems, have assumed they were debris from our own satellites.

The small comets probably have their origin in a disk of cometary material located beyond the orbit of Neptune. I have proposed that these objects are sent streaming into the inner solar system by the passage of an as-yet-undiscovered Dark Planet through the outer regions of this immense disk. But several times every 26 million years, the orbit of this planet crosses the inner portion of this disk, where the known large comets are located. This triggers a storm of large comets Earth-ward, eventually causing widespread destruction and the loss of many species. Such an event may well be responsible for the death of the dinosaurs 65 million years ago.

The small comets, meanwhile, have been slowly changing the face of the rest of the solar system as well. These objects may also be causing the mysterious bright flashes of light seen on the Moon, the presence of water vapor in the atmosphere of Venus, the periods of flowing water on Mars, the heating of the atmosphere of the outer planets, and the icy composition of their distant moons. I am certain that in the years to come we will find many other traces of the effects of the small comets both here on Earth and elsewhere in the solar system.

Looking back, I now think that the discovery of the small comets was inevitable. After nearly three decades of space exploration, scientists have begun in just the last few years to aim their satellites, their eyes in space, towards the Earth. Our home is finally getting the attention it deserves. If we had not dis-covered the evidence for the small comets with the imager on the Dynamics Explorer satellite, someone else with another camera like it would have done so sooner or later. Ours was just the first of its kind to fly on an Earth satellite. The future will see many more satellites geared exclusively to the study of the Earth and its environment in space. There will be other surprises. The hole in the ozone layer was one surprise. The small comets were another. This is just the beginning.

The timing was right for this discovery to be made. The technology was available. Someone had only to use it and we did just that. Yet, despite our use of state-of-the-art instruments, the small comet observations took place at the limits of detection. This is not a place to shy away from. This is the frontier where many scientific discoveries are made. It is where new particles are found in physics, where unsuspected lifeforms are discovered in the deepest seas, where the clues to earthquake prediction are being sought. The satellite images of the atmospheric holes, the measurements of water bursts in our atmosphere, and the telescope observations of the small comets were all made at this frontier.

If these images, measurements, and observations do not represent the small comets—as some still insist—then what are they? Why should the rate of the unidentified streaks in Yeates's pictures be the same as the rate of the atmospheric holes found by Dynamics Explorer if they are not identical objects? Why should the atmospheric holes in the Dynamics Explorer images resemble those in the images from the Viking satellite? Why should they be of the same size and number? And what were Olivero and his colleagues recording with their microwave antenna in Pennsylvania, if not the bursts of water vapor from the small comets? You would need to propose something even more fantastic than the small comets themselves to account for all of these things, and what that might be, I just do not know.

I am now as confident of the reality of the small comets as I can be without holding one in my hands. Only a sliver of doubt remains, a one-in-a-thousand chance or less that the small comets are not real. I believe they do exist, but I am too cautious to say that I am certain of it. Certainty is as elusive in science as it is in life. It is a comfort only for the narrow-minded and the dead.

But because people are certain that the Earth is isolated from the universe, I think it may take another five, ten, or more years before the idea of small comets is finally accepted. I have done my part. People will no longer listen to me on this subject, as science does not take too kindly to those who plow through its well-tended gardens. Most likely, the process of accepting these objects into the scientific fold will fall to the new generation of scientists, for as the physicist Max Planck once drolly remarked: Scientists don't change their minds, they just die. Today's young scientists will be responsible for changing our mistaken impressions about the role of these objects in our past, present, and future.

This brings me to the end of my account and to a final note on final things. The number of small comets, if not their size, tends to strike a note of terror in people. In this sense, at least, the small comets are the latest chapter in a long book. From the earliest times, large comets have been regarded as objects of fear and superstition. They were seen as symbols and harbingers of wars and famines and other great natural and human disasters. Some even saw them as signs of retribution by the gods for the sins of men. This view led one 16th century rationalist to quip: "If comets were caused by the sins of mortals, they would never be absent from the sky."

In keeping, then, with proper tradition, I will end with a brief tale of cometary peril. If the present influx of small comets continues—and there is no reason to think that it is likely to end anytime soon—then four billion years from now, when the Earth's age has doubled, our planet will have twice the amount of water it has today. All during this time there will be a little evaporation from the top of our atmosphere but it will never keep up with the thousand times greater rainfall from the small comets. The Earth will just continue to accumulate water.

The doubling of the water on the Earth will no doubt affect the delicate balance that we live in. But we do not know what this will mean for us. Will we just have twice the water and less land? Or will this be enough water to turn the Earth into an ice planet, or a complete water planet? All we know is that something is going to happen. Fortunately, it is not going to happen next year.

It will be a very gradual filling-up, like the steady drip of a faucet into a barrel. There is not going to be any severely rapid change from the fall of one extra inch of cosmic rain on Earth every 10,000 years. We are safe. We will not have an ocean beach here in Iowa for a while yet.

Notes

1   A Radical Departure

A radical departure. L.A. Frank, J.B. Sigwarth, and J.D. Craven, 1986. "On the Influx of Small Comets Into the Earth's Upper Atmosphere, II. Interpretation." *Geophysical Research Letters*, 13: 307 (April).

"Astronomers should leave to astrologers..." Anonymous, 1985. "Miscasting the Dinosaur's Horoscope." *The New York Times*, April 2.

Ongoing planet formation. Nils Aall Barricelli, 1985. "Preferential Pehrihelion and Aphelion Distances, and Planetary Formation." *Theoretic Papers* Vol.3 No.6, Oslo.

Large comets exhaustively studied. Carl Sagan and Ann Druyan, 1985. *Comet*. New York, Random House.

Previous work on small comets. Herbert A. Zook, J.A. Fernández and E. Grün, 1985. "Selection Effects Against Small Comets." In R.H. Giese and P. Lamy, eds. *Properties and Interactions of Interplanetary Dust*. D. Reidel, p. 287.

2   Cosmic Rain

One story bore the mark. Philip M. Boffey, 1986. "Did Ocean Waters Arrive in Rain of Icy Comets?" *The New York Times*, April 1.

A pair of papers. L.A. Frank, J.B. Sigwarth, and J.D. Craven, 1986. "On the Influx of Small Comets Into the Earth's Upper Atmosphere, I. Observations." *Geophysical Research Letters*, 13: 303 (April). L.A. Frank, J.B. Sigwarth, and J.D. Craven, 1986. "On the Influx of Small Comets Into the Earth's Upper Atmosphere, II. Interpretation." *Geophysical Research Letters*, 13: 307 (April).

Gleeful critics. Mark Washburn, 1988. "The Waters Above, the Storm Below." *Sky & Telescope*, December, p. 628.

"Crazy as they come..." Patrick Huyghe, 1986. "Origin of the Ocean." *Oceans*, July-August, p. 10.

"I've never heard anything like it..." *Ibid.*

"There is nothing like..." *Ibid.*

The circuit. Gina Kolata, 1990. "To Make the Big Time of Science, Better Take Your Show on the Road." *The New York Times*, January 2.

3   The Black Spot Mystery

It all began in 1981. L.A. Frank and J.D. Craven, 1988. "Imaging Results from Dynamics Explorer 1." *Reviews of Geophysics*, 26: 249 (May).

"How can we get rid..." James Ehmann, 1986. "Cosmic Comets of the Sea." *Omni*, July, p. 72.

In public for the first time. J.B. Sigwarth, L.A. Frank, J.D. Craven, 1983. "Atmospheric Holes Possibly Associated with Meteors." *EOS*, May 3, p. 281, abstract.

Press picks up story. Stefi Weisburd, 1985. "Atmospheric Footprints of Icy Meteors." *Science News*, Dec. 21 & 28, p. 391.

Fourth presentation. J.B. Sigwarth, L.A. Frank, J.D. Craven, 1985. "Latitudinal and Longitudinal Distribution of Atmospheric Holes Associated with Meteors." *EOS*, November 12, p. 1005, abstract.

Description and interpretation papers. L.A. Frank, J.B. Sigwarth, and J.D. Craven, 1986. "On the Influx of Small Comets Into the Earth's Upper Atmosphere, I. Observations." *Geophysical Research Letters*, 13: 303 (April). L.A. Frank, J.B. Sigwarth, and J.D. Craven, 1986. "On the Influx of Small Comets Into the Earth's Upper Atmosphere, II. Interpretation." *Geophysical Research Letters*, 13: 307 (April).

"Uncomfortable ramifications." A.J. Dessler letter to L.A. Frank, Feb. 7, 1986.

"If you restrict the journal... ." A.J. Dessler interview by P. Huyghe, Oct. 14, 1987.

4   True Confessions

Co-author of two scientific papers. James A. Van Allen and Louis A. Frank, 1959. "Survey of Radiation Around the Earth to a Radial Distance of 107,400 Kilometers." *Nature*, 183: 430. James A. Van Allen and Louis A. Frank, 1959. "Radiation Measurements to 658,300 Kilometers with Pioneer IV." *Nature*, 184: 219.

Hannes Alfvén. John Noble Wilford, 1989. "Novel Theory Challenges Big Bang." *The New York Times*, Feb. 23.

Hannes Alfvén. Stephen G. Brush, 1990. "Prediction and Theory Evaluation, Alfvén on Space Plasma Phenomena." *EOS*, Jan. 9., p. 19.

Alfred Wegener. Martin Schwarzbach, 1986. *Alfred Wegener: The Father of Continental Drift*. Madison, Wisconsin, Science Tech.

5 A Storm of Controversy

"A terse distillation..." A.J. Dessler, 1986. "A Turbulent Interface." *Geophysical Research Letters*, 13: 1 (January).

Comments received. A.J. Dessler letter to L.A. Frank, August 13, 1986.

First Comment. Thomas M. Donahue, 1986. "Comment on the Paper 'On the Influx of Small Comets Into the Earth's Upper Atmosphere, II. Interpretation'." *Geophysical Research Letters*, 13: 555 (June).

"Pretty unavoidable." Philip M. Boffey, 1986. "Did Ocean Waters Arrive in Rain of Icy Comets?" *The New York Times*, April 1.

Angry letter. T.M. Donahue letter to J. A. Van Allen, April 1, 1986.

"Bitter." Stephen Cain, 1988. "Comet Theory Data Misused, U-M Man Says." *Ann Arbor News*, May 18.

Donahue denies it. T.M. Donahue letter to L.A. Frank, May 24, 1988.

"Most important finding in space science." Philip M. Boffey, *Ibid*.

"We regret the public contention ... " Unpublished letter to *The New York Times* from T.M. Donahue and L.A. Frank, April 22, 1986.

"A large number of problems." Thomas M. Donahue, 1986. *Ibid*, p. 555.

How Venus lost its oceans. Thomas M. Donahue, J.H. Hoffman, R.R. Hodges, Jr., and A.J. Watson, 1982. "Venus Was Wet: A Measurement of the Ratio of Deuterium to Hydrogen." *Science*, 216: 630 (May 7).

My Reply to Donahue. L.A. Frank, J.B. Sigwarth, and J.D. Craven, 1986. "Reply [to Donahue]." *Geophysical Research Letters*, 13: 559 (June).

Others had questioned water on Venus. Sig J. Bauer, 1983. "Water on Venus: Lack or Loss?" *Annales Geophysicae*, 1: 477.

"Your observations and interpretations..." S.J. Bauer letter to L.A. Frank, May 28, 1986.

"Venus's present hydrogen loss rate..." John S. Lewis, 1974. "Volatile Element Influx on Venus from Cometary Impacts." *Earth and Planetary Science Letters*, 22: 239.

Microwave measurements. John Olivero, J.J. Tsou, C.L. Hale, and R.G. Joiner, 1986. "Solar Absorption Microwave Measurement of the Upper Atmospheric Water Vapor." *Geophysical Research Letters*, 13: 197.

"Ingenious." D. Hunten letter to L.A. Frank, April 14, 1986.

6 The Creation of the Oceans

Bathtubs full of water. Fred Powledge, 1982. *Water*. New York, Farrar Straus Giroux, p. 23.

Alternate view. Carl Sagan and Ann Druyan, 1985. *Comet*. New York, Random House, p. 317.

"Trickle down theory..." Patrick Huyghe, 1986. "The Origin of the Oceans." *Oceans*, July-August, p. 11.

Lack of evidence to the contrary. *Ibid*. p. 12.

"Data concerning ocean volume..." Thomas M. Donahue, 1986. "Comment on the Paper 'On the Influx of Small Comets into the Earth's Upper Atmosphere II. Interpretation'." *Geophysical Research Letters*, 13: 555 (June).

"The exceedingly slow rate..." William W. Rubey, 1951. "Geologic History of Seawater: An Attempt to State the Problem." *Bulletin of the Geological Society of America*, 62: 1135 (September).

"By small increments..." *Ibid.* p. 1117.

Problem of excess volatiles not solved. P. Wilde letter to L.A. Frank, June 29, 1988.

"Some of the excess water..." P. Wilde interview by P. Huyghe, Nov. 29, 1989.

"A startling geological revelation..." P. Wilde letter, *Ibid.*

A study of Halley's Comet. D. Krankowsky, P. Lämmerzahl, I. Herrwerth, J. Woweris, P. Eberhardt, U. Dolder, U. Herrmann, W. Schulte, J.J. Berthelier, J.M. Illiano, R.R. Hodges and J.H. Hoffman, 1986. "*In Situ* Gas and Ion Measurements at Comet Halley." *Nature*, 321: 326 (May 15).

Halley's fingerprint. P. Eberhardt, U. Dolder, W. Schulte, D. Krankowsky, P. Lämmerzahl, J. H. Hoffman, R.R. Hodges, J.J. Berthelier and J.M. Illiano, 1987. "The D/H Ratio in Water from Comet P/Halley." *Astronomy and Astrophysics*, 187: 435. P. Eberhardt, D. Krankowsky, W. Schulte, U. Dolder, P. Lämmerzahl, J.J. Berthelier, J. Woweries, U. Stubbemann, R.R. Hodges, J.H. Hoffman, and J.M. Illiano, 1987. "On the CO_2 and N_2 Abundance in Comet P/Halley." *Astronomy and Astrophysics*, 187: 481.

7 Heat, Dust, and the Origin of Life

Comet watchers flock to Alice Springs. Peter H. Lewis, 1986. "Disappointed Comet-Gazers Bidding Halley's Farewell." *The New York Times*, April 13.

Critic on composition. David Parry Rubincam, 1986. "Comment on the Paper 'On the Influx of Small Comets into the Earth's Upper Atmosphere, II. Interpretation'." *Geophysical Research Letters*, 13: 701 (July).

Critic on composition. Christopher P. McKay, 1986. "Comment [on the Paper 'On the Influx of Small Comets into the Earth's Upper Atmosphere, II. Interpretation']." *Geophysical Research Letters*, 13: 976 (September).

Critic on composition. Donald E. Morris, 1986. "Comment on 'On the Influx of Small Comets into the Earth's Upper Atmosphere, II. Interpretation'." *Geophysical Research Letters*, 13: 1482 (December).

Critics on composition. John T. Wasson and Frank T. Kyte, 1987. "Comment on the Letter 'On the Influx of Small Comets into the Earth's Upper Atmosphere, II. Interpretation'." *Geophysical Research Letters*, 14: 779 (July).

My Replies. L.A. Frank, J.B. Sigwarth, and J.D. Craven, 1986. "Reply [to Rubincam]." *Geophysical Research Letters*, 13: 703 (July). L.A. Frank, J.B. Sigwarth, and J.D. Craven, 1986. "Reply [to McKay]." *Geophysical Research Letters*, 13: 979 (September). L.A. Frank, J.B. Sigwarth, and J.D. Craven, 1986. "Reply to Morris." *Geophysical Research Letters*, 13: 1484 (December). L.A.

Frank, J.B. Sigwarth, and J.D. Craven, 1987. "Reply to Wasson and Kyte." *Geophysical Research Letters*, 14: 781 (July).

Laboratory experiments. L.J. Lanzerotti, W.L. Brown and R.E. Johnson, 1985. "Laboratory Studies of Ion Irradiations of Water, Sulfur Dioxide, and Methane Ices." In J. Klinger, D. Benest, A. Dollfus, R. Smoluchowski, eds. *Ices in the Solar System*. Boston, D. Reidel. R.E. Johnson, J.F. Cooper, L.J. Lanzerotti, G. Strazzulla, 1987. "Radiation Formation of a Non-Volatile Crust." *Astronomy and Astrophysics*, 187: 889.

Halley's Comet. Hans Balsiger et al., 1986. "Ion Composition and Dynamics at Comet Halley." *Nature*, 321: 330.

Carbon mantle. Donald E. Morris, 1986. "Comment on 'On the Influx of Small Comets into the Earth's Upper Atmosphere, II. Interpretation'." *Geophysical Research Letters*, 13: 1482 (December).

Examination of carbon on Earth. Heinrich D. Holland, 1984. *The Chemical Evolution of the Atmosphere and Oceans*. Princeton, Princeton University Press, p. 82.

Biochemical molecules from comets. J. Oró, 1961. "Comets and the Formation of Biochemical Compounds on the Primitive Earth." *Nature*, 190: 389.

Study of comets and life. William Ward Maggs, 1989. "Comets and Life." *EOS*, March 28.

8   The Atmosphere and the Ice Ages

Noctilucent clouds. George C. Reid, 1975. "Ice Clouds at the Summer Polar Mesopause." *Journal of Atmospheric Science*, 32: 523. M. Gadsden, 1982. "Noctilucent Clouds." *Space Science Reviews*, 33: 280.

"Greatly exceed those that have been observed." George C. Reid and Susan Solomon, 1986. "On the Existence of an Extraterrestrial Source of Water Vapor in the Middle Atmosphere." *Geophysical Research Letters*, 13: 1129.

Solomon's earlier point. Susan Solomon et al., 1982. "On the Chemistry of H_2O, H_2 and Meteoritic Ions in the Mesosphere and Lower Thermosphere." *Planetary and Space Science*, 30: 1117.

Results from intercosmos satellites. R. Knuth, G. Sonnemann, D. Felske, L. Martini and B. Stark, 1976. "Some Problems and Results of Solar Occultation Measurements in the Thermosphere. Part II: Neutral Gas Density Variations." *COSPAR Space Research XVI*, Berlin, Akademie-Verlag. D. Felske, R. Knuth, G. Sonnemann, L. Martini and B. Stark, 1977. "On the Additional Lyman Alpha Absorber in the Winter-Thermosphere." *Journal of Atmospheric and Terrestrial Physics*, 39: 1423. G. Sonnemann, D. Felske, R. Knuth, L. Martini and B. Stark, 1977. "How Dry is the Thermosphere?" *COSPAR Space Research XVII*, New York, Pergamon Press.

Rocket-borne measurements of water vapor. K.U. Grossmann, W.G. Frings, D. Offermann, L. André, E. Kopp, and D. Krankowsky, 1985. "Concentrations of H_2O and NO in the Mesosphere and the Lower Thermosphere at High Latitudes." *Journal of Atmospheric and Terrestrial Physics*, 47: 291.

Three launches, three positive results. K.U. Grossmann interview by P. Huyghe, Oct. 14, 1987.

Water vapor concentrations in the middle atmosphere. J.J. Olivero, J.J. Tsou, C.L. Hale, and R.G. Joiner, 1986. "Solar Absorption Microwave Measurement of the Upper Atmospheric Water Vapor." *Geophysical Research Letters*, 13: 197.

"We thought Frank's hypothesis..." J. Olivero interview by P. Huyghe, Oct. 13, 1987.

Master's thesis. Dennis M. Adams, 1988. "Extreme Short Term Variability in Upper Atmospheric Water Vapor as Measured by Ground-Based Microwave Radiometry." Master's Thesis, Pennsylvania State University, University Park, August.

Olivero's paper. Dennis M. Adams, John J. Olivero, and Charles L. Croskey, 1987. "A Search for Extra-Terrestrial Water in the Upper Atmosphere by Ground-Based Microwave Radiometry." *EOS*, 68: 372 (April 21), abstract.

"Observed, short-term perturbations..." D.M. Adams, *Ibid* p. 90.

Not let down. J.J. Olivero letter to L.A. Frank, March 2, 1990.

Four years of data analyzed. M.F. Bonadonna, J.J. Olivero and C.L. Croskey, 1990. "In Search of Small Comets: H_2O Bursts Observed in the Mesosphere." *EOS*, 71: 570 (April 24), abstract.

9   A Masquerade for Radar

"If his comets have all this dust..." Patrick Huyghe, 1986. "Origin of the Ocean." *Oceans*, July-August, p. 12.

"Our radar is infrared..." *Ibid*.

"NORAD has a grouping of radars..." *Ibid*.

"A target strength of enough ice..." *Ibid*.

Several billion meteors. Malcolm W. Browne, 1989. "Radio System Uses Fiery Meteor Trails to Transmit Data." *The New York Times*, Aug. 22.

A study of radar meteor rates. E.L. Vogan and L.L. Campbell, 1957. "Meteor Signal Rates Observed in Forward-Scatter." *Canadian Journal of Physics*, 35: 1176.

Paper prepared in June 1986. L.A. Frank, J.B. Sigwarth, and J.D. Craven, 1986. "Temporal Variations of the Influxes of Small Comets Into the Earth's Atmosphere." University of Iowa Research Report, 86-33.

Dessler was inflexible. A.J. Dessler letter to L.A. Frank, Aug. 6, 1986.

Meteor trails enlisted by Pentagon. Malcolm W. Browne, *Ibid*.

Experiment of meteor burst communications. Billy P. Ficklin, 1986. "Possible Comets." SRI International Working Paper, Project 8055, July 8.

Small comet search by radar. Gérard Caudal, 1989. "EISCAT Finds No Small Comet." Centre de Recherche en Physique de l'Environnement, CNET-CNRS, Paris.

10 How to Spot a Small Comet

Öpik's Tables. Ernst J. Öpik, 1963. "Tables of Meteor Luminosities." *Irish Astronomy Journal*, 6: 3.

We had our answer. L.A. Frank, J.B. Sigwarth, and J.D. Craven, 1987. "Estimate of the Luminosity of the Impact of Small Comets with the Earth's Atmosphere." University of Iowa Research Report, 87-18.

Dustballs. Ernst J. Öpik, 1958. *Physics of Meteor Flight in the Atmosphere*. New York, Interscience.

11 Flying Saucers and Other Strange Events

"This meteor is not characteristic..." J. West letter to L.A. Frank, Aug. 21, 1986.

"Unlike a trailed meteor..." E. Kurczewski letter to L.A. Frank, April 5, 1986.

Nebulous meteors. William Corliss, ed., 1986. *The Sun and Solar System Debris*. Glen Arm, Maryland, Sourcebook Project, p. 232.

"Direct evidence." K. Bigg letter to L.A. Frank, Dec. 8, 1986.

Bigg's account. E.K. Bigg and W.J. Thompson, 1969. "Daytime Photograph of a Group of Meteor Trails." *Nature*, 222: 156 (April 12).

Mystery object photographed. Bruce Maccabee, 1987. "Analysis of an Anomalous Image Found in a DSMP Weather Satellite Photograph." Fund for UFO Research, Mt. Rainier, Maryland.

"The most likely explanation yet" for UFOs. A.H. Lawson letter to L.A. Frank, July 12, 1986.

"There is water..." Charles Fort, 1974. *The Complete Books of Charles Fort*. New York, Dover Publications, p. 180 and p. 188.

Study of ice falls. James E. McDonald, 1960. "The Ice-Fall Problem." *Weatherwise*, June, p. 110.

Griffiths' icefall. Simon Welfare, 1980. *Arthur C. Clarke's Mysterious World*. New York, A & W, p. 40.

"Ties in beautifully." A.C. Clarke letter to K. Keeton, cc L.A. Frank, June 26, 1986.

Chinese ice falls. Hu Zhong-Wei letter to L.A. Frank, April 24, 1989.

Tunguska. John D. O'Keefe and Thomas J. Ahrens, 1982. "Cometary and Meteorite Swarm Impact on Planetary Surfaces." *Journal of Geophysical Research*, 87: 6669 (August 10).

Flynn Creek. *Ibid.* p. 6668.

Bright flashes of light. William Corliss, ed., 1982. *Lightning, Auroras, Nocturnal Lights and Related Luminous Phenomena*. Glen Arm, Maryland, Sourcebook Project, p. 180.

Odd halos and strange rainbows. William Corliss, ed., 1984. *Rare Halos, Mirages, Anomalous Rainbows, and Related Electromagnetic Phenomena*. Glen Arm, Maryland, Sourcebook Project, p. 6.

Expanding ball of light. Richard F. Haines, 1988. "Expanding Ball of Light (EBL) Phenomenon." *Journal of Scientific Exploration*, 2: 83.

Clear Air Turbulence. H. Blake letter to A.J. Dessler, cc L.A. Frank, April 15, 1986.

A rare event. Frank W. Gibson, 1976. "A Rare Event in the Stratosphere." *Nature*, 263: 487 (Oct. 7).

Sodium clouds. T.J. Beatty, R.E. Bills, K.H. Kwon and C.S. Gardner, 1988. "CEDAR Lidar Observations of Sporadic Na Layers at Urbana, IL." *Geophysical Research Letters*, 15: 1137 (September).

12   A Scramble for Satellite Data

OGO-4. Robert R. Meier and Talbot A. Chubb, 1987. "Theoretical and Observational Problems with 'Holes' in the Far UV Dayglow." Naval Research Laboratory Report 9078, November 5.

Dynamics Explorer 2. William B. Hanson, 1986. "Comment [on the Paper 'On the Influx of Small Comets into the Earth's Upper Atmosphere, II. Interpretation']." *Geophysical Research Letters*, 13: 981 (September).

Re-examination of data. L.A. Frank, J.B. Sigwarth, and J.D. Craven, 1986. "Reply [to Hanson]." *Geophysical Research Letters*, 13: 985 (September).

A satellite called Viking. C.D. Anger, S.K. Babey, A.L. Broadfoot, R.G. Brown, L.L. Cogger, R. Gattinger, J.W. Haslett, R.A. King, D.J. McEwen, J.S. Murphree, E.H. Richardson, B.R. Sandel, K. Smith and A. Vallance Jones, 1987. "An Ultraviolet Auroral Imager for the Viking Spacecraft." *Geophysical Research Letters*, 14: 387.

Joint paper. L.A. Frank, J.B. Sigwarth, J.D. Craven, J.S. Murphree and L.L. Cogger, 1988. "A Search for Atmospheric Holes in Viking Images of Earth's Ultraviolet Dayglow." *EOS*, April 19, p. 413, abstract.

Viking confirms findings. L.A. Frank, J.B. Sigwarth and J.D. Craven, 1989. "Search for Atmospheric Holes with the Viking Cameras." *Geophysical Research Letters*, 16: 1457 (December).

13   Extraordinary Evidence

Size of atmospheric hole as function of altitude. Talbot A. Chubb, 1986. "Comment on the Paper 'On the Influx of Small Comets into the Earth's Upper Atmosphere, I. Observations'." *Geophysical Research Letters*, 13: 1075 (October).

Our Reply. L.A. Frank, J.B. Sigwarth, and J.D. Craven, 1986. "Reply [to Chubb]." *Geophysical Research Letters*, 13: 1079 (October).

Revised Comment. B.L. Cragin, W.B. Hanson, R.R. Hodges, and D. Zuccaro, 1987. "Comment on the Papers 'On the Influx of Small Comets into the Earth's Upper Atmosphere, I. Observations and II. Interpretation'." *Geophysical Research Letters*, 14: 573 (May).

When we did the same analysis. L.A. Frank, J.B. Sigwarth, and J.D. Craven, 1987. "Reply to Cragin et al." *Geophysical Research Letters*, 14: 557 (May).

Master's thesis. John B. Sigwarth, 1988. "Imaging of Absorption and Emission Features of Water-Vapor Clouds Associated with Small Comets as Observed with Dynamics Explorer 1." Master's Thesis, University of Iowa, Iowa City, May.

Sigwarth presentation. J.B. Sigwarth, L.A. Frank and J.D. Craven, 1988. "Imaging of Absorption and Emission Features of Water-Vapor Clouds Associated with Small Comets as observed with Dynamics Explorer 1." *EOS*, 69: 1350 (Nov. 1), abstract.

14 Launch of the Artificial Comets

Third launch. M. Mendillo et al., 1989. "Preliminary Results from ERIC-3: Attempts to Create Atmospheric Signatures of Comet-Like Objects." *EOS*, 70: 405 (April 11), abstract. Mark Washburn, 1988. "The Waters Above, the Storm Below." *Sky & Telescope*, Dec., p. 628.

Space shuttle water-vapor. M. Mendillo, J. Baumgardner, D.P. Allen, J. Foster, J. Holt, G.R.A. Ellis, A. Klekociuk and G. Reber, 1987. "Spacelab-2 Plasma Depletion Experiments for Ionospheric and Radio Astronomical Studies." *Science*, 238: 1260.

First launch. M. Mendillo interview by P. Huyghe, Dec. 3, 1987.

Polar BEAR. Anonymous, 1986. "Museum-Piece Satellite Goes Into Space." *Science News*, 130: 361, Dec. 6.

Second launch. Michael Mendillo, John B. Sigwarth, John D. Craven, Louis A. Frank, John Holt and David Tetenbaum, 1990. "Project ERIC: The Search for Environmental Reactions Induced by Comets." *Advances in Space Research*, 10: (7)83.

15 Some Cometary Competition

Donahue's last shot. Thomas M. Donahue, 1987. "Small Comets: Implications for Interplanetary Lyman Alpha." *Geophysical Research Letters*, 14: 213.

Discrepancies in Lyman alpha. Ralph C. Bohlin, 1973. "Mariner 9 Ultraviolet Spectrometer Experiment: Measurements of the Lyman Alpha Sky Background." *Astronomy and Astrophysics*, 28: 323.

Donahue's cometesimals. Thomas M. Donahue, Tamas I. Gombosi and Bill R. Sandel, 1987. "Cometesimals in the Inner Solar System." *Nature*, 330: 548 (Dec. 10).

His baby comets. Patrick Huyghe, 1988. "Oceans from Comets—New Evidence." *Oceans*, April, p. 9.

Donahue's detailed paper for *Icarus*. Thomas M. Donahue, Tamas I. Gombosi and Bill R. Sandel, 1987. "Evidence for Cometesimals in the Inner

Solar System: Interplanetary Lyman Alpha and Lunar Cratering." University of Michigan.

Wetherhill's reaction. G. W. Wetherhill letter to J.A. Burns, January 28, 1988.

"I think that Donahue's work..." A.J. Dessler interview by P. Huyghe, Oct. 14, 1987.

Hydrogen gas around Halley's. J.M. Ajello, A.I. Stewart, G.E. Thomas and A. Graps, 1987. "Solar Cycle Study of Interplanetary Lyman-Alpha Variations: Pioneer Venus Orbiter Sky Background Results." The Astrophysical Journal, 317: 964.

Dynamics Explorer looks at Halley's. J.D. Craven et al., 1986. "The Hydrogen Coma of Comet Halley Before Perihelion: Preliminary Observations with Dynamics Explorer 1." Geophysical Research Letters, 13: 873.

"But who's right?" T.M. Donahue letter to L.A. Frank, Dec. 23, 1987.

16 The Verdict

Donahue cometesimals. Thomas M. Donahue, Tamas I. Gombosi and Bill R. Sandel, 1987. "Cometesimals in the Inner Solar System." Nature, 330: 548 (Dec. 10).

Published in Nature. James A. Van Allen and Louis A. Frank, 1959. "Survey of Radiation Around the Earth to a Radial Distance of 107,400 Kilometers." Nature, 183: 430. James A. Van Allen and Louis A. Frank, 1959. "Radiation Measurements to 658,300 Kilometers with Pioneer IV." Nature, 184: 219.

My interpretation of Donahue's Voyager results. L.A. Frank, J.B. Sigwarth, and J.D. Craven, 1988. "An Atomic Hydrogen Torus Around the Sun from a Large Population of Small Comets." University of Iowa Research Report, 88-9.

"At this stage..." D. Lindley letter to L.A. Frank, Feb. 22, 1988.

News and views. Paul D. Feldman, 1987. "Encounters of the Second Kind." Nature, 330: 518 (Dec. 10).

Three stories. Richard A. Kerr, 1988. "In Search of Elusive Little Comets." Science, 240: 1403 (June 10). Richard A. Kerr, 1988. "Comets Were A Clerical Error." Science, 241: 532 (July 29). Richard A. Kerr, 1989. "Double Exposures Reveal Mini-Comets." Science, 243: 170 (Jan. 13).

17 The Competition Fizzles

Torus of hydrogen. L.A. Frank, J.B. Sigwarth, and J.D. Craven, 1988. "An Atomic Hydrogen Torus Around the Sun from a Large Population of Small Comets." University of Iowa Research Report, 88-9.

Galileo spacecraft. John Noble Wilford, 1987. "Swarms of Miniature Comets May Elude Detection." The New York Times, Dec. 10.

Bombshell. D.T. Hall and D.E. Shemansky, 1988. "No Cometesimals in the Inner Solar System." *Nature*, 335: 417 (Sept. 29).

"The primary objective…" Donald Hunten, 1988. "The Hunten Report." The University of Arizona.

"I blew it." Richard A. Kerr, 1988. "Comets Were A Clerical Error." *Science*, 241: 532 (July 29).

Final version. D.T. Hall and D.E. Shemansky. *Ibid.*

Mistake snowballed. Richard A. Kerr, *Ibid.*

18   Why the Moon Doesn't Ring Like a Bell

One response on the Moon problem. R.B. Baldwin, 1987. "On the Current Rate of Formation of Impact Craters of Varying Sizes on the Earth and Moon." *Geophysical Research Letters*, 14: 216.

Making the same mistake. P.M. Davis, 1986. "Comment on the Letter 'On the Influx of Small Comets into the Earth's Upper Atmosphere'." *Geophysical Research Letters*, 13: 1181 (Nov.)

"To ring like a bell." *Ibid.*

The most difficult problem. L.A. Frank, J.B. Sigwarth, and J.D. Craven, 1986. "Reply to Davis and Nakamura et al." *Geophysical Research Letters*, 13: 1186 (November).

Dessler wrote me. A.J. Dessler letter to L.A. Frank, Sept. 26, 1986.

Nakamura and Oberst Comment. Yosio Nakamura, Jürgen Oberst, Stephen M. Clifford, and Bruce G. Bills, 1986. "Comment on the Letter 'On the Influx of Small Comets into the Earth's Upper Atmosphere, II. Interpretation'." *Geophysical Research Letters*, 13: 1184 (November).

Those who worked on the physics of lunar impacts. Gary V. Latham, Maurice Ewing, Frank Press, George Sutton, James Dorman, Yosio Nakamura, Nafi Toksoz, David Lammlein, and Fred Duennebier, 1972. "Passive Seismic Experiment." *Apollo 16 Preliminary Science Report*, NASA SP-315, p. 9-1.

My Reply to the Moon problem. L.A. Frank, J.B. Sigwarth, and J.D. Craven, *Ibid.*

Brilliant work. Yakov B. Zel'dovich and Yu.P. Raizer, 1967. In W. D. Hayes and R.F. Probstein, eds. *Physics of Shock Waves and High-Temperature Hydrodynamic Phenomena*, Vol. II. New York, Academic Press.

"Splattered into the Moon…" Mark Washburn, 1988. "The Waters Above, the Storm Below." *Sky & Telescope*, Dec. p. 629.

"Most interesting…" J. Oberst letter to A.J. Dessler, cc L.A. Frank, Sept 28, 1986.

"A flat-floored crater…" John D. O'Keefe and Thomas J. Ahrens, 1982. "Cometary and Meteorite Swarm Impact on Planetary Surfaces." *Journal of Geophysical Research*, 87: 6668 (Aug. 10).

Bright flash on Moon. Gary V. Latham et al., *Ibid.* p. 9-1.

History of strange glows on Moon. William Corliss, ed., 1985. *The Moon and the Planets*. Glen Arm, Maryland, Sourcebook Project, p. 126.

Sigwarth runs in with article. G. Kolovos, J.H. Seiradakis, H. Varvoglis, and S. Avgoloupis, 1988. "Photographic Evidence of a Short Duration: Strong Flash from the Surface of the Moon." *Icarus*, 76: 525 (December).

"From a point slightly..." *Ibid.* p. 528.

Uruguayan astronomer. Julio A. Fernández, 1988. "End-States of Short-Period Comets and their Role in Maintaining the Zodiacal Dust Cloud." *Earth, Moon, and Planets*, 41: 155.

"To the actual absence..." *Ibid.* p. 159.

Ice trapped at lunar poles. James R. Arnold, 1979. "Ice in the Lunar Polar Regions." *Journal of Geophysical Research*, 84: 5659 (Sept. 10).

Burst of water vapor reported. J.W. Freeman Jr., H.K. Hills, R.A. Lindeman, and R.R. Vondrak, 1973. "Observations of Water Vapor Ions at the Lunar Surface." *The Moon*, 8: 115.

"Firmly ruled out ..." *Ibid.* p. 125.

19   Mars and the Outer Planets

"These networks resemble..." Harold Masursky, J.M. Boyce, A.L. Dial, G.G. Schaber, and M.E. Strobell, 1977. "Classification and Time of Formation of Martian Channels Based on Viking Data." *Journal of Geophysical Research*, 82: 4022 (September 30).

"May have formed..." *Ibid.* p. 4027.

"The water ice deposits..." H.H. Kieffer, S.C. Chase, T.Z. Martin, E.D. Miner, and F.D. Palluconi, 1976. "Martian North Polar Summer Temperature: Dirty Water Ice." *Science*, 194: 1343 (Dec. 17).

Cyclical Martian climate. B.M. Jakosky and E.S. Barker, 1984. "Comparison of Ground-Based and Viking Orbiter Measurements of Martian Water Vapor: Variability of the Seasonal Cycle." *Icarus*, 57: 322.

Warm period on Mars. Jonathan Eberhart, 1989. "Signs of Old Mars: Written in the Dust." *Science News*, 135: 173 (March 18).

Disposing of water on Mars. Fraser P. Fanale, James R. Salvail, Aaron P. Zent, and Susan E. Postawko, 1986. "Global Distribution and Migration of Subsurface Ice on Mars." *Icarus*, 67: 1.

Similar thoughts. S.H. Zisk and P.J. Mouginis-Mark, 1980. "Anomalous Region on Mars: Implications for Near-Surface Liquid Water." *Nature*, 288: 126 (Nov. 13).

Water ice structures. Jonathan Eberhart, 1989. "Neptune: A New Page in the Book of Worlds." *Science News*, 136: 300 (Nov. 4).

Planetary rings. John Noble Wilford, 1989. "New Clues Emerge in Mystery of Planetary Rings." *The New York Times*, June 27.

Heating of the outer planets. L.A. Frank, J.B. Sigwarth, and J.D. Craven, 1986. "On the Influx of Small Comets into the Earth's Upper Atmosphere, II. Interpretation." *Geophysical Research Letters*, 13: 309 (April).

Atmospheric temperatures of Jupiter, Saturn, Uranus, and Neptune. R.A. Hanel, B.J. Conrath, L.W. Herath, V.G. Kunde and J.A. Pirraglia, 1981. "Albedo, Internal Heat, and Energy Balance of Jupiter: Preliminary Results of the Voyager Infrared Investigation." *Journal of Geophysical Research*, 86: 8705. R.A. Hanel, B.J. Conrath, V.G. Kunde, J.C. Pearl, and J.A Pirraglia, 1983. "Albedo, Internal Heat Flux, and Energy Balance of Saturn." *Icarus*, 53: 262. H. Moseley, B. Conrath and R.F. Silverberg, 1985. "Atmospheric Temperature Profiles of Uranus and Neptune." *The Astrophysical Journal*, 292: L83.

20   Where the Small Comets Come From

Oort's paper. Jan. H. Oort, 1950. "The Structure of the Cloud of Comets Surrounding the Solar System, and a Hypothesis Concerning Its Origin." *Bulletin of the Astronomical Institutes of the Netherlands*, 11: 91 (Jan. 13).

Oort Cloud supplied by Oort Disk. J.G. Hills, 1981. "Comet Showers and the Steady-State Infall of Comets from the Oort Cloud." *The Astronomical Journal*, 86: 1730 (November).

Photograph of Beta Pictoris. Bradford A. Smith and Richard J. Terrile, 1984. "A Circumstellar Disk Around Beta Pictoris." *Science*, 226: 1421 (Dec. 21).

Likely origin of large comets. Martin Duncan, Thomas Quinn and Scott Tremaine, 1988. "The Origin of Short-Period Comets." *The Astrophysical Journal*, 328: L69 (May 15).

Likely candidate for the task. L.A. Frank, J.B. Sigwarth, and J.D. Craven, 1986. "Reply to Morris." *Geophysical Research Letters*, 13: 1484 (December). L.A. Frank, 1989. "Atmospheric Holes and the Small Comet Hypothesis." *Australian Physicist*, 26: 19 (January-February).

Work of physicists. Edgar Everhart, 1968. "Change in Total Energy of Comets Passing through the Solar System." *The Astronomical Journal*, 73: 1039 (December). Edgar Everhart, 1969. "Close Encounters of Comets and Planets." *The Astronomical Journal*, 74: 735 (June). Julio A. Fernández, 1980. "On the Existence of a Comet Belt Beyond Neptune." *Monthly Notices of the Royal Astronomical Society*, 192: 481.

21   Death of the Dinosaurs

Alvarez scenario. Luis W. Alvarez, W. Alvarez, F. Asaro and H.V. Michel, 1980. "Extraterrestrial Cause for the Cretaceous-Tertiary Extinction." *Science*, 208: 1095. Luis W. Alvarez, 1987. "Mass Extinctions Caused by Large Bolide Impacts." *Physics Today*, 40: 24 (July).

Similar periodicity in work of paleontologists. David M. Raup and John J. Sepkoski, Jr., 1986. "Periodic Extinction of Families and Genera." *Science*, 231: 833 (Feb. 21).

Additional evidence. M.R. Rampino and R.B. Stothers, 1984. "Terrestrial Mass Extinctions, Cometary Impacts and the Sun's Motion Perpendicular to the Galactic Plane." *Nature*, 308: 709.

Nemesis. Richard Muller, 1988. *Nemesis: The Death Star.* New York, Weidenfeld & Nicolson.

Planet X. Daniel P. Whitmire and Albert A. Jackson IV, 1984. "Are Periodic Mass Extinctions Driven by a Distant Solar Companion?" *Nature*, 308: 713. D.P. Whitmire and J.J. Matese, 1985. "Periodic Comet Showers and Planet X." *Nature*, 313: 36. J.J. Matese and D.P. Whitmire, 1986. "Planet X and the Origins of the Shower and the Steady State Flux of Short-Period Comets." *Icarus*, 65: 37.

Dark Planet. L.A. Frank, J.B. Sigwarth, and J.D. Craven, 1988. "A Hypothesis Concerning the Inner Oort Disk and Its Relationship to Comet Showers and Extinction of Species." University of Iowa Research Report, 88-8.

Carbon at geological boundary. W.S. Wolbach, R.S. Lewis, and E. Anders, 1985. "Cretaceous Extinctions: Evidence for Wildfires and Search for Meteoritic Material." *Science*, 230: 167.

Shocked quartz. B.F. Bohor, P.J. Modreski and E.E. Foord, 1987. "Shocked Quartz in the Cretaceous-Tertiary Boundary Clays: Evidence for a Global Distribution." *Science*, 236: 705.

Ages of large craters on Earth. W. Alvarez and R.A. Muller, 1984. "Evidence from Crater Ages for Periodic Impacts on the Earth." *Nature*, 308: 718.

Paper not well received. L.A. Frank et al. *Ibid.*

22   The Debate Comes to an End

"The importance for scientific progress..." A.J. Dessler, 1986. "Controversial Publications: The Role of Comments and Replies." *Geophysical Research Letters*, 13: 1363 (December).

Paper on cometesimals. Thomas M. Donahue, Tamas I. Gombosi, and Bill R. Sandel, 1987. "Cometesimals in the Inner Solar System." *Nature*, 330: 548.

"Scientists would like to think..." Patrick Huyghe, 1988. "Oceans from Comets—New Evidence." *Oceans*, April, p. 11.

"Only if the small-comet hypothesis..." A.J. Dessler letter to L.A. Frank, Dec. 5, 1986.

Last Comment. John T. Wasson and Frank T. Kyte, 1987. "Comment on the Letter 'On the Influx of Small Comets into the Earth's Upper Atmosphere, II. Interpretation'." *Geophysical Research Letters*, 14: 779 (July).

Comets coated with viruses. Fred Hoyle and Chandra Wickramasinghe, 1979. *Diseases from Space.* New York, Harper & Row. Fred Hoyle and Chandra Wickramasinghe, 1981. *Evolution From Space.* New York, Simon & Schuster. Fred Hoyle and Chandra Wickramasinghe, 1985. *Living Comets.* Cardiff, University College Cardiff Press.

"A nice, appropriate response." A.J. Dessler letter to L.A. Frank, Jan. 27, 1987.

"As I see it..." A.J. Dessler letter to L.A. Frank, March 24, 1987.

Explaining to Dessler. L.A. Frank letter to A.J. Dessler, April 6, 1987.

Poem for Dessler's birthday. L.A. Frank letter to A.J. Dessler, Oct. 24, 1988.

"I'm still convinced..." Tex (A.J. Dessler) letter to L.A. Frank, April 5, 1989.

"Dear Mr. Excitement..." Tex (A.J. Dessler) letter to L.A. Frank, April 20, 1989.

23 Astronomers Enter the Fray

Soter's Comment. Steven Soter, 1987. "Comment on the Paper 'On the Influx of Small Comets into the Earth's Upper Atmosphere.'" *Geophysical Research Letters*, 14: 162 (February).

GEODSS Network. Jesse Hall, 1989. "Deep Space Surveillance Keeps Track of Objects From Gloves to Satellites." *Space Trace*, December, p. 6.

Taft papers. Laurence G. Taft, 1981. "A New Asteroid Observation and Search Technique." *Publication of the Astronomical Society of the Pacific*, 93: 658. Laurence G. Taff, 1986. "Satellite Debris: Recent Measurements." *Journal of Spacecraft*, 23: 342 (May-June).

"Frank is wrong..." L.G. Taft interview by P. Huyghe, Oct. 15, 1987.

Time variations in occurrence of atmospheric holes. L.A. Frank, J.B. Sigwarth, and J.D. Craven, 1986. "Temporal Variations of the Influxes of Small Comets Into the Earth's Atmosphere," University of Iowa Research Report, 86-33.

I redid Soter's work. L.A. Frank, J.B. Sigwarth, and J.D. Craven, 1987. "Reply to Soter." *Geophysical Research Letters*, 14: 164 (February).

24 Where Are You Now, Galileo?

"I could more easily believe..." Maurice Ebison, ed., 1977. *The Harvest of a Quiet Eye*. London, The Institute of Physics.

Gehrels' review. T. Gehrels letter to L.A. Frank, July 17, 1986.

Soter convinced small comets should have been seen. Steven Soter, 1987. "Comment on the Paper 'On the Influx of Small Comets into the Earth's Upper Atmosphere.'" *Geophysical Research Letters*, 14: 162 (February).

Gehrels' paper. Tom Gehrels, 1985. "Asteroids and Comets." *Physics Today*, 38(2): 33.

"Several times the energy..." T. Gehrels letter *Ibid*.

Gehrels may have detected them. Tom Gehrels and R.S. McMillan, 1982. "CCD Scanning for Asteroids and Comets." In W. Fricke and G. Teleki, eds. *Sun and Planetary Systems*. Dordrecht, Holland, Reidel. T. Gehrels, B.G.

Marsden, R.S. McMillan and J.V. Scotti, 1986. "Astrometry with a Scanning CCD." *Journal of Astronomy*, 91: 1242. T. Gehrels and F. Vilas, 1986. "A CCD Search for Geosynchronous Debris." *Icarus*, 68: 412.

25   The Great Search Begins

"I thought it was pretty wild..." C.M. Yeates interview by P. Huyghe, Nov. 14, 1989.

"We were sitting..." *Ibid*.

"A lot of astronomers..." Patrick Huyghe, 1988. "Oceans from Comets—New Evidence." *Oceans*, April, p. 11.

Yeates search method. C.M. Yeates, 1989. "Initial Findings from a Telescopic Search for Small Comets Near Earth." *Planetary and Space Science*, 37: 1185 (October).

"I went through my own calculation..." E.M. Shoemaker letter to A.J. Dessler, cc C.M. Yeates, Jan. 4, 1987 (sic).

"The failure to find Frank's microcomets..." *Ibid*.

"It was a pretty crude system..." C.M. Yeates interview by P. Huyghe, Nov. 14, 1989.

The runs on the telescope. C.M. Yeates letter to T. Gehrels, July 19, 1988.

Unidentified streaks appear in 36 images. C.M. Yeates, 1989. *Planetary and Space Science, Ibid*.

"The whole thing..." C.M. Yeates interview, 1989. *Ibid*.

26   Double Trouble

"I was shocked..." C.M. Yeates interview by P. Huyghe, Nov. 14, 1989.

"Gehrels was telling the press..." *Ibid*.

"Apparently confirmed..." Jet Propulsion Laboratory Public Information Office Release #1191, March 30, 1988.

Yeates's findings made public. Patrick Huyghe, 1988. "Oceans from Comets—New Evidence." *Oceans*, April, p. 11. Richard Monastersky, 1988. "Comet Controversy Caught on Film." *Science News*, 133: 340 (May 28).

"What I was left with..." C.M. Yeates, *Ibid*.

Consecutive images first suggested. C.M. Yeates letter to T. Gehrels, July 19, 1988.

"If you see them..." T. Gehrels note to C.M. Yeates, April, 1988.

Analysis of consecutive images. L.A. Frank, J.B. Sigwarth, and C.M. Yeates, 1990. "A Search for Small Solar-System Bodies Near the Earth Using a Ground-Based Telescope: Technique and Observations." *Astronomy and Astrophysics*, 228:522 (February).

Short summary. L.A. Frank, J.B. Sigwarth, J.D. Craven, C.M. Yeates, and T. Gehrels, 1988. "Telescopic Search for Small Comets in Consecutive Images with the Spacewatch Camera." *EOS*, 69: 1293, Nov. 1, abstract.

"If only you could..." T. Gehrels letter to C.M. Yeates, July 11, 1988.

"We have been in this business..." T. Gehrels letter to L.A. Frank, Oct. 17, 1988.

My reply. L.A. Frank letter to T. Gehrels, Oct. 25, 1988.

"The complete absence..." C.M. Yeates letter to T. Gehrels, Nov. 2, 1988.

"Your reasonings are impressive..." T. Gehrels letter to C.M. Yeates, Nov. 14, 1988.

"Relax..." T. Gehrels postcard to L.A. Frank, Nov. 2, 1988.

"If there indeed were as many cometesimals..." T. Gehrels to B. Jakosky, cc L.A. Frank, Nov. 14, 1988.

"We express our appreciation..." L.A. Frank, J.B. Sigwarth, J.D. Craven, and C.M. Yeates, 1988. "Telescopic Search for Small Comets in Consecutive Images with the Spacewatch Camera." AGU Poster paper, December.

"The normal criteria for acceptance..." E.M. Shoemaker letter to A.J. Dessler, cc: C.M. Yeates, 1988.

"For your paper..." A.J. Dessler letter to C.M. Yeates, Sept. 9, 1988.

"*Utterly impossible...*" C.M. Yeates letter to A.J. Dessler, Sept. 28, 1988.

"Well, it wouldn't have mattered..." C.M. Yeates interview by P. Huyghe, Nov. 14, 1989.

Scientific American. John Horgan, 1989. "A Snowball's Chance." *Scientific American*, March, p.21.

Run-in with Dessler over a paper. B.R. Clemesha, P.P. Batista and D.M. Simonich, 1988. "Concerning the Origin of Enhanced Sodium Layers." *Geophysical Research Letters*, 15: 1267 (October).

Keeping small comets out of the literature. B.R. Clemesha letter to L.A. Frank, March 5, 1990.

No one believed it. Richard A. Kerr, 1989. "Double Exposures Reveal Mini-Comets?" *Science*, 243: 170 (Jan. 13).

Poster paper on double images. L.A. Frank, J.B. Sigwarth, J.D. Craven, and C.M. Yeates, 1988. *Ibid.*

27   The Turning Point

Warren E. Leary, 1989. "Big Asteroid Passes Near Earth Unseen in a Rare Close Call." *The New York Times*, April 20.

Yeates's paper published. C.M. Yeates, 1989. "Initial Findings from a Telescopic Search for Small Comets Near Earth." *Planetary and Space Science*, 37: 1185 (October).

Consecutive image paper published. L.A. Frank, J.B. Sigwarth, and C.M. Yeates, 1990. "A Search for Small Solar-System Bodies Near the Earth Using a Ground-Based Telescope: Technique and Observations." *Astronomy and Astrophysics*, 228:522 (February).

Long paper about small comets. A.J. Dessler, 1989. "The Small Comet Hypothesis." Submitted to *Reviews of Geophysics*, Oct.

On atmospheric holes in Viking images. L.A. Frank, J.B. Sigwarth, and J.D. Craven, 1989. "Search for Atmospheric Holes with the Viking Cameras." *Geophysical Research Letters*, 16: 1457 (December).

Work of Scott Bolton. S.J. Bolton, 1990. "One Year Variations in the Near Earth Solar Wind Ion Density and Bulk Flow Velocity." *Geophysical Research Letters*, 17: 37 (January).

Observatory in Iowa. John Horgan, 1989. "A Snowball's Chance." *Scientific American*, March, p.21.

Novel use of new technology. Peter M. Banks, 1989. "A New Means for Observation of Small Comets and Other Water-Laden Bodies Entering Earth's Upper Atmosphere." *Geophysical Research Letters*, 16: 575 (June).

Mende proposal. S.B. Mende, 1989. "Technical Notes Concerning a Search for Cometary Water Vapor Clouds Near Earth with the Space Shuttle." Lockheed Palo Alto Research Laboratories.

28   Small Comets and the Future

Quip by 16th century rationalist. Carl Sagan and Ann Druyan, 1985. *Comet*. New York, Random House, p. 29.

Index